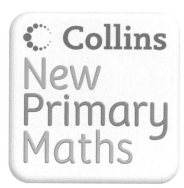

Collins
New
Primary
Maths

Pupil Book 6B

Series Editor: Peter Clarke

Authors: Jeanette Mumford, Sandra Roberts, Andrew Edmondson

Contents

Page number

Unit D2

Unit E2

Ordering decimals

1 Read these decimal fractions. What does the red digit represent?

  a 2·56   b 3·58   c 1.09   d 7·67   e 4·84

2 Order each group of decimal fractions from smallest to largest.

  a 5·62, 4·89, 3·72, 1·63, 2·04   d 1·51, 1·82, 1·55, 1·87, 1·59

  b 4·86, 4·93, 4·12, 4·51, 4·66   e 4·21, 2·41, 4·12, 2·14, 2·22

  c 0·72, 0·78, 0·75, 0·79, 0·74   f 31·78, 31·08, 31·8, 31·87

3 Draw a number line and write the hundredths that come between these tenths.

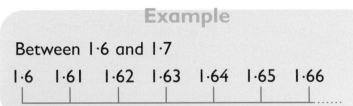

Example

Between 1·6 and 1·7

1·6   1·61   1·62   1·63   1·64   1·65   1·66

  a
    4·2   4·3
  b
    2·7   2·8
  c
    5·5   5·6
  d
    3·3   3·4
  e
    7·1   7·2

1 Read these decimal fractions. What does the red digit represent?

  a 6·872   d 2·004

  b 5·314   e 8·068

  c 3·129   f 1·507

2 Order each group of decimal fractions from smallest to largest.

a 4·687, 4·293, 4·106, 4·005, 4·972

b 3·612, 5·892, 4·637, 2·637, 1·073

c 0·832, 0·045, 1·405, 2·832, 1·504

d 5·310, 4·624, 5·130, 0·513, 0·462

e 6·723, 6·581, 6·729, 6·735, 6·526

3 Draw a number line and write the thousandths that come between these hundredths.

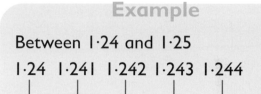

Example

Between 1·24 and 1·25

1·24 1·241 1·242 1·243 1·244

a
4·67   4·68

b
2·89   2·90

c
3·31   3·32

d
5·04   5·05

e
6·53   6·54

f
7·22   7·23

1 Write out these number sequences.

a From 0 to 6 in jumps of 0·6.

b From 0 to 6 in jumps of 0·3.

c From 10 to 17 in jumps of 0·7.

Decimal decisions

- Round a decimal with two decimal places to the nearest tenth or to the nearest whole number
- Order decimals with up to three places

 1 What whole numbers are these decimals between?
Underline the number the decimal is closest to.

a | ? | 4·8 | ?

b | ? | 5·6 | ?

Example

| <u>3</u> | 3·4 | 4 |

c | ? | 2·3 | ?

d | ? | 1·7 | ?

e | ? | 9·2 | ?

f | ? | 4·56 | ?

g | ? | 9·7 | ?

h | ? | 3·98 | ?

i | ? | 7·49 | ?

j | ? | 6·52 | ?

2 Order these decimals from smallest to largest.

a 8·3, 8·6, 8·1, 8·7, 8·9, 8·2

b 2·26, 2·45, 2·78, 2·12, 2·95, 2·06

c 5·62, 5·26, 5·66, 5·22, 5·06, 5·02

d 7·94, 7·96, 7·49, 7·07, 7·69, 7·99

e 3·06, 3·03, 3·63, 3·36, 3·13, 3·31

f 9·5, 9·56, 9·6, 9·15, 9·67, 9·7

g 2·6, 2·59, 2·8, 2·14, 2·87, 2·9

h 4·74, 4·47, 4·7, 4·4, 4·77, 4·89

i 6·73, 6·51, 6·4, 6·82, 6·1, 6·9

j 12.81, 12·8, 12·63, 12·5, 12·77, 12·48

1 Copy this table and fill in the tenths and whole numbers that the number in the middle is between. Circle the tenth and whole number it is closest to.

whole number	tenths	number	tenths	whole number
4	(4·6)	4·63	4·7	(5)
		7·92		
		3·49		
		0·51		
		8·75		
		9·16		
		22·06		
		45·99		
		31·39		
		17·28		
		34·72		
		58·51		

2 Order these decimals from smallest to largest.

a 5·23, 5·2, 5·36, 5·6, 5·32, 5·61

b 7·68, 7·66, 7·8, 7·86, 7·866, 7·066

c 2·001, 2·021, 2·21, 2·211, 2·202, 2·2

d 9·75, 9·57, 9·5, 9·771, 9·571, 9·7

e 21·4, 12·46, 21·406, 12·64, 12·4, 21·401

f 47·5, 47·05, 47·005, 47·15, 47·475, 47·015

g 84·1, 84·41, 84·441, 84·114, 84·4, 84·14

h 66·96, 66·99, 66·996, 66·9, 66·696, 66·6

i 0·412, 0·4, 0·21, 0·42, 0·224, 0·2

j 15·02, 15·26, 15·026, 15·6, 15·226, 15·262

I am thinking of a number…

a If I count on 12 tenths I get to 4·8. What number am I thinking of?

b If I count on 25 hundredths I get to 7·62. What number am I thinking of?

c If I count on 30 hundredths I get to 10·21. What number am I thinking of?

d If I count on 13 thousandths I get to 5·004. What number am I thinking of?

e If I count on 24 thousandths I get to 13·473. What number am I thinking of?

Pay your share!

● Solve problems involving fractions

You need:

● a calculator

These groups of friends went out for a pizza.
They only ordered **one** pizza to share.
Work out how much each person has to pay.

Show your working out.

Number of people	Price per pizza
4	£8
8	£8
3	£9.30
6	£9.30
6	£6.30
12	£6.60

Check your answer on a calculator.

Some friends go out for a pizza together. They have to work out how much each person has to pay when the bill comes. Work out how much each person pays in each group.

You need:

● a calculator

Show your working out.

Number of Pizzas	Number of people	Price per pizza
3	4	£7.60
6	8	£8
4	3	£9.30
4	6	£9.30
5	6	£6.30
8	12	£6.60
12	18	£9.90

Check your answer on a calculator.

Look at your answers for the ● section.

What fraction of the bill did each person pay in each question?

Can you work out what this is as a percentage?

Written methods for addition and subtraction

Choose whether to add or subtract each pair of numbers.

You should do about half addition and half subtraction calculations.

a	651	241	**i**	439	624
b	274	481	**j**	704	284
c	183	562	**k**	89·3	65·1
d	406	381	**l**	35·5	26·3
e	373	583	**m**	163·8	165·1
f	841	385	**n**	45·25	26·74
g	872	392	**o**	254·56	231·27
h	275	638	**p**	634·21	401·76

Explain what you find easy or difficult about using this method.

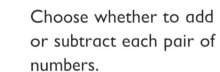

Choose whether to add or subtract each pair of numbers.

You should do about half addition and half subtraction calculations.

You need:
● calculator

❶ a	3673	5259	**d**	9536	3227
b	7109	4283	**e**	6294	5017
c	5271	2740	**f**	12 341	16 371

g 15 269   24 812   n 41 612   7947

h 28 513   31 451   o 234·73   85·41

i 33 418   28 366   p 69·34   248·37

j 41 781   39 023   q 281·06   351·82

k 5692   24 936   r 581·23   203·31

l 37 348   9067   s 2863·5   428·33

m 28 651   8744   t 639·35   2542·41

2 Now check your answers with a calculator. If you made any mistakes can you work out where you went wrong? Explain what you have learnt from your mistakes.

3 What do you have to remember when you use a written method for decimals?

Write out instructions for using a written method either for addition or subtraction.

Give your instructions to a friend to test them out!

School problems

● **Solve multi-step problems**

Work out the problems. Record the calculations and say how you worked them out: in your head, in your head with jottings, using the written method or with a calculator. Be sure to make an estimate first and check your answers.

Rosehill Infant School has 162 children.

a There are 48 children in Reception and 56 in Year 1. How many children is that?

b The rest of the children are in Year 2. How many children are there in Year 2?

c $\frac{1}{2}$ of the children live less than 500 metres from the school. How many children is that?

d All the children bring gloves to school one day. How many gloves are there in the school?

e It costs £4.30 to go on a trip. How much will it cost for three children?

Ingleton Primary School has 57 nursery children, 193 Key Stage 1 children and 248 Key Stage 2 children.

1 a How many children are there in Key Stage 1 and Key Stage 2?

b If 22 children are away, how many children will be at school altogether?

c In key stage 1, 48 children have 1 brother or sister, 87 have 2 brothers or sisters and the rest have none. How many children have no brothers or sisters?

d School dinners cost £3.40. In one family there are 3 children who all have school dinner. How much do they pay each day?

e $\frac{1}{4}$ of Key Stage 2 children come to school by bus. How many children is that?

f $\frac{1}{8}$ of Key Stage 2 children bring packed lunches. How many children is that?

g A sixth of the school were absent with the flu one week. How many children was that?

h The whole school is going on a trip. Each coach can hold 72 people. How many coaches will the school need to book if 74 adults go as well?

2 Make up two word problems about Ingleton School for a friend to work out.

a When Mum and her two children, Rose and Daisy, got on the scales, it showed 127 kg. When Daisy got off, it showed 103 kg. When Rose got off and Daisy got back on, it showed 89 kg. How much did Mum and her two children weigh?

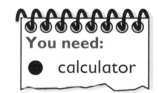

You need:
● calculator

b Anthony spent £38.20. He bought a scarf and three hats. The scarf cost £11.95. How much was each hat?

c Theo and Luke both start reading the same book on the same day. Theo reads 6 pages a day and Luke reads 9. What page will Luke be on when Theo is on page 72? The book has 288 pages altogether. How many days will Theo take to read the book?

In your head

Partition each of these calculations to find the answer.

a 38 × 6

b 45 × 7

c 53 × 4

d 66 × 5

e 87 × 8

f 73 × 8

g 96 × 4

h 63 × 9

i 94 × 7

j 46 × 7

k 57 × 6

l 83 × 9

m 64 × 8

Example

$27 \times 8 = (20 \times 8) + (7 \times 8)$
$= 160 + 56$
$= 216$

1 Copy each grid. Fill in the answers on the grid by multiplying the numbers in the horizontal line by the number in the circle, and multiplying the numbers in the vertical line by the number in the square.

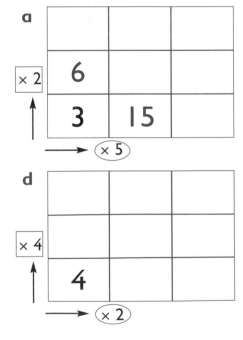

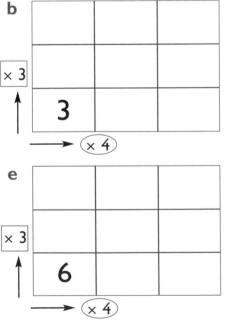

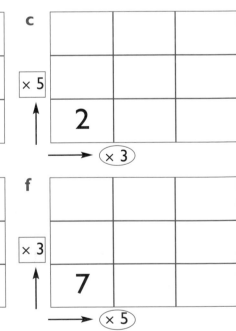

2 Find the missing number by multiplying mentally.

a 56p × ☐ = £3.36

b 37p × ☐ = £3.33

c 29p × ☐ = £2.03

d 84p × ☐ = £6.72

e 46p × ☐ = £2.76

f 72p × ☐ = £5.76

g 63p × ☐ = £4.41

h ☐ × 95 = 380

i ☐ × 4 = 312

j ☐ × 39 = 195

k ☐ × 74 = 666

l ☐ × 3 = 279

m ☐ × 67 = 536

n ☐ × 8 = 616

1 2 3 4 5 6 7 8 9

For each target number below, choose any three numbers from above once.

Arrange the numbers like this

    ×

to make a product of:

a 44·5

b 44·8

c 45·0

d 60·8

e 53·6

f 63·9

Multiply the answer mentally, then record all the calculations.

Multiplication methods

● **Use written methods for HTU × TU**

 Choose five calculations. Approximate the answer first, then use the grid method to work out the answer.

a
176×13

b
149×25

c
234×32

d
194×37

e
245×14

f
376×12

g
246×28

h
319×45

i
462×38

Example

$235 \times 12 \approx 200 \times 12 = 2400$

×	200	30	5	
10	2000	300	50	2350
2	400	60	10	+ 470

$$\frac{2820}{1}$$

1 Write multiplication facts for each set of numbers.

Example

$9 \times 30 = 270$

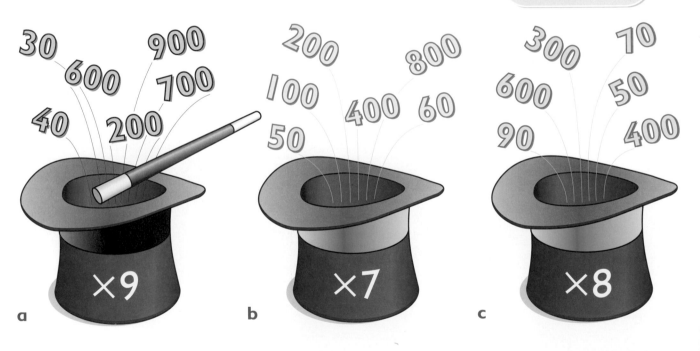

a ×9 **b** ×7 **c** ×8

2 Approximate the answer first. Use the expanded method of recording to work out the answer to each calculation.

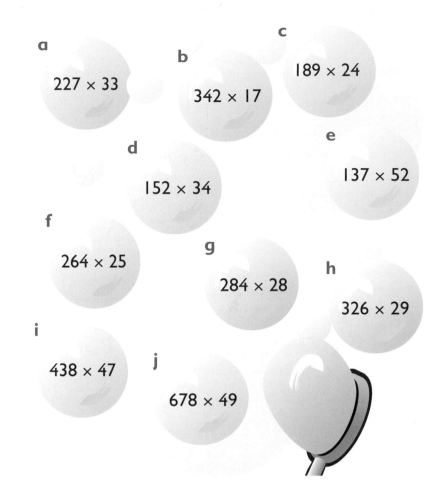

a 227 × 33

b 342 × 17

c 189 × 24

d 152 × 34

e 137 × 52

f 264 × 25

g 284 × 28

h 326 × 29

i 438 × 47

j 678 × 49

Example

246 × 23 ≈ 250 × 20 = 5000

```
    246
  ×  23
   4000    200 × 20
    800     40 × 20
    120      6 × 20
    600    200 × 3
    120     40 × 3
     18      6 × 3
   5658
```

1 Use each of these digits once.

Arrange them to make a product as close as possible to

10 000

You need:

● calculator

2 Use each of these digits once.

Arrange them to make a product as close as possible to

10 000

Efficient multiplication methods

Partition each of these calculations in
2 ways and work out the answers.

a
156 × 37

b
228 × 15

Example

325 × 42 → 300 × 42 or 325 × 40
            20 × 42   325 × 2
             5 × 42

c
345 × 24

d
196 × 33

e
429 × 26

f
484 × 63

g
748 × 53

h
927 × 86

① Approximate the answer first.

② Find the answer to each calculation. Use the efficient written method.

a
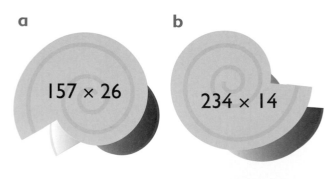
157 × 26

b
234 × 14

Example

246 × 23 ≈ 250 × 20 = 5000

```
     246
   ×  23
    4920    246 × 20
     738    246 ×  3
    5658
      1
```

c
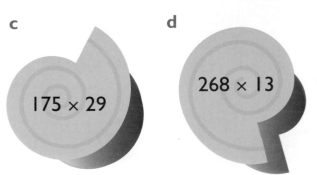
175 × 29

d
268 × 13

Remember

Remember to keep the digits
in the correct columns.

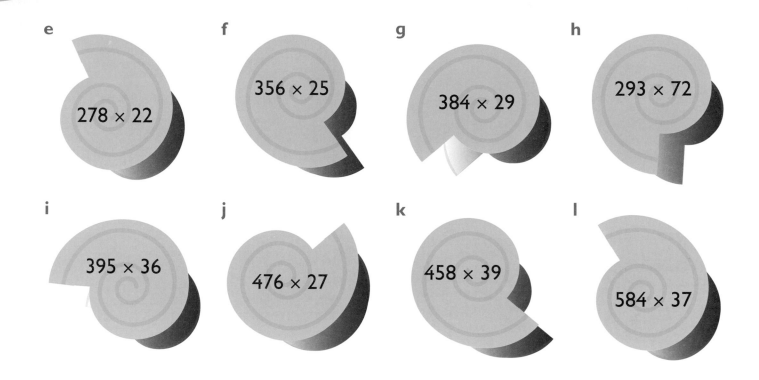

e 278 × 22

f 356 × 25

g 384 × 29

h 293 × 72

i 395 × 36

j 476 × 27

k 458 × 39

l 584 × 37

1 Make up at least 3 word problems that involve calculations in the form of

☐ ☐ ☐ × ☐ ☐

using ideas from the picture.

2 Ask your partner to solve your problems.

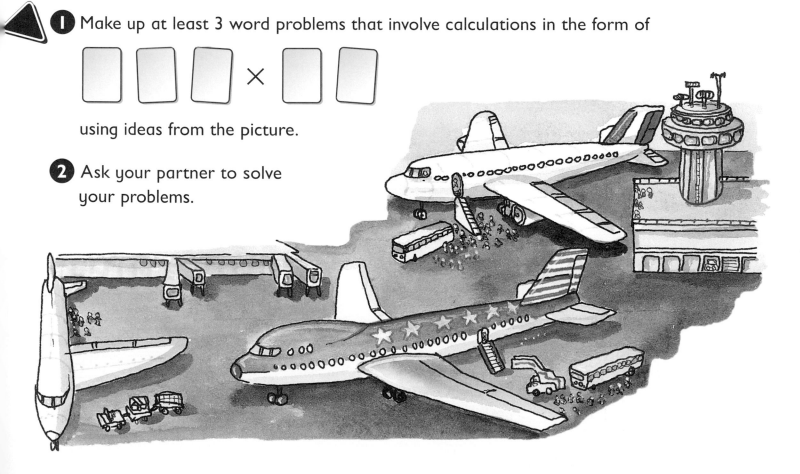

Computer word problems

● **Choose and use appropriate strategies to solve problems**

 Use the items and prices from the ● activity to work out the answers to these questions. You will not need to include VAT on any of the items.

a The shop sells 7 computers in one day. How much money do they make?

b 2 lap-tops are sold each day, Monday to Saturday. How much money does the shop make?

c The shop has 56 digital cameras in stock. What is their total value?

d The shop has 100 lap-tops to sell. How much money will they make?

e The shop is planning a sale. All items above £500 will be reduced by 20%. What is the new price of each item?

f In the sale, all items below £500 will be reduced by 15%. What is the new price of each item?

 Read the word problems **a** to **e**.

Choose an appropriate method of calculating your answer.

- ● mental
- ● mental with jottings
- ● paper and pencil (efficient method)
- ● calculator

You need:
- ● calculator

Prices are exclusive of VAT at 17·5%

£1020 £380 £40 £260

£850

a Work out the total cost of each item including VAT at 17·5%.

b Buy the computer and the printer as a package and get 20% off the total price. How much do you pay including VAT?

c Buy 1 of each item. What is the total cost before VAT? What is the cost after VAT is added?

d The shop is offering a 25% discount if you buy 2 mobile phones. What is the total cost including VAT?

e The shop is offering £5 of free calls with every mobile phone sold. Calls cost 10p per minute peak time and 5p per minute off-peak. How many minutes can you speak for free in peak times or off-peak times?

f In the sale, the lap-top will be reduced in price by 10% before VAT is added. What will be the new price excluding VAT? How much will you pay, including VAT?

Use the prices and the items in the ● activity to work out the answers.

A local business buys 2 of each of the items.

1 Calculate the total cost including VAT at 17·5%.

2 They spread the payments over a 12-month period at an extra cost of 25% interest. What is the new total cost? What is the monthly payment?

You need:
● calculator

Patterns in multiplication

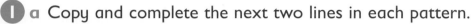

1 a Copy and complete the next two lines in each pattern.

Row 1	$9 \times 1 + 1 =$	Row 1	$0.9 \times 1 + 0.1 =$
Row 2	$9 \times 2 + 2 =$	Row 2	$0.9 \times 2 + 1.2 =$
Row 3	$9 \times 3 + 3 =$	Row 3	$0.9 \times 3 + 2.3 =$
Row 4		Row 4	
Row 5		Row 5	

b Predict the answer to the Row 7 of each pattern.
Check your predictions.

2 Work out the missing digits and complete the calculations.

a
$$\begin{array}{r} 3\square \cdot 2 \\ \times\ 6 \\ \hline \square 05 \cdot \square \end{array}$$

b
$$\begin{array}{r} \square 8 \cdot 5 \\ \times\ \square \\ \hline 13\square \cdot 0 \end{array}$$

c
$$\begin{array}{r} 89 \cdot 5 \\ \times\ \square \\ \hline 7\square\square \cdot 0 \end{array}$$

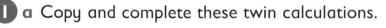

1 a Copy and complete these twin calculations.

Use your calculator.

i 12×42 and 21×24

ii 12×84 and 21×48

iii 13×62 and 31×26

iv 23×96 and 48×46

v 24×63 and 42×36

b Write about the results
you found.

You need:

● calculator

2 **a** Copy and complete the pattern:

   Use your calculator.

   b Look for a pattern.

   Use it to predict the answers to:

   i 8 × 123 456 + 6

   ii 8 × 12 345 678 + 8

> 8 × 1 + 1 =
>
> 8 × 12 + 2 =
>
> 8 × 123 + 3 =
>
> 8 × 1234 + 4 =
>
> 8 × 12 345 + 5 =

3 Work out the missing digits without using your calculator.

   a ☐7 × ☐ = 102   **b** 3☐ × ☐ = 170

   c ☐6 × ☐4 = 544   **d** ☐9 × 4☐ = 1392

4 **a** Find the answer to 321 × 54.   Use a calculator.

   b Use the same 5 digits ☐ ☐ ☐ × ☐ ☐
      so that the product is:

   i less than 5 000

   ii more than 15 000

   iii between 10 000 and 15 000

Copy and complete.

> 99 × 11 =
>
> 99 × 22 =
>
> 99 × 33 =
>
> 99 × 44 =

You need:

● calculator

Extend the pattern to 99 × 99.

Compare the first and last answers and write what you notice.

What if you added 11 to 99 and multiplied by 99?

Number sequences

Write the next 10 numbers in each of these sequences using the rule shown.

a The rule is: add 0·2 each time.

1,   1·2,   1·4, ...

+ 0·2   + 0·2   + 0·2

b The rule is: add 0·1 each time.

6·5,   6·6, ...

+ 0·1   + 0·1

c The rule is: add 0·5 each time.

7·5,   8·0, ...

+ 0·5   + 0·5

d The rule is: add 0·25 each time.

3·5,   3·75, ...

+ 0·25   + 0·25

e The rule is: subtract 0·5 each time.

10·5,   10·0, ...

− 0·5   − 0·5

f The rule is: subtract 0·25 each time.

9·5,   9·25 ....

− 0·25   − 0·25

---

1 Add 0·5 to each of these.

a	0·5	f	6·5
b	0·2	g	3·4
c	1·5	h	5·2
d	0·6	i	0·9
e	2·0	j	12·0

2 Add 0·2 to these.

a	0·2	f	0·6
b	1·0	g	0·1
c	0·5	h	1·7
d	1·3	i	6·5
e	0·8	j	9·0

3 Add 0·25 to these.

a	0·25	f	0·75
b	0·5	g	2·0
c	2·25	h	0
d	1·0	i	3·75
e	4·0	j	5·5

4 Only the middle number is given in these sequences.

- Choose a rule from the box.
- Copy and complete your sequence.
- Choose a different rule each time.

a ☐ ☐ ☐ 16 ☐ ☐ ☐

b ☐ ☐ ☐ 7 ☐ ☐ ☐

c ☐ ☐ ☐ 11·1 ☐ ☐ ☐

d ☐ ☐ ☐ 0·5 ☐ ☐ ☐

e ☐ ☐ ☐ 10 ☐ ☐ ☐

f ☐ ☐ ☐ −0·75 ☐ ☐ ☐

Rules

+ 0·2

− 0·1

+ 2·5

− 0·2

+ 0·5

+ 0·25

− 0·5

5 When you have finished question **4**, swap your six number sequences with a friend. What is the rule for each of their number sequences?

What to do

- Choose any starting number.
- Choose one of the following types of number sequences:
  - **i** Add or subtract the same number
  - **ii** Multiply or divide by the same number each time
  - **iii** Add or subtract a changing number
  - **iv** Add the previous two numbers
  - **v** Combine two operations
- Write the first ten numbers in the sequence.
- Repeat five times.
- Then swap with a friend. Can they identify the rule and write the next three numbers in the sequence?

Deriving related facts

Use table facts to work out related facts with decimals

 1 Pick 2 cards, one from the blue barrel and one from the red barrel.

- Do what the blue card tells you.
- A winning score lies between 0 and 10.

How many different winning scores can you make?

2 Copy and complete.

a $2.6 \times 10 = \boxed{}$

b $9.3 \times \boxed{} = 930$

c $\boxed{} \times 10 = 37$

d $126 \div \boxed{} = 1.26$

e $\boxed{} \div 100 = 52$

f $0.8 \div \boxed{} = 0.08$

 1 Write 3 facts you can derive from each of these statements. At least one of the facts must include decimals.

a $40 \times 60 = 2400$

b $19 \times 90 = 1710$

c $25 \times 70 = 1750$

d $16 \times 16 = 256$

e $850 \div 50 = 17$

f $96 \div 6 = 16$

g $300 \times 7 = 2100$

h $60 \times 24 = 1440$

i $1080 \div 40 = 27$

j $4000 \div 80 = 50$

k $3420 \div 9 = 380$

l $1240 \div 40 = 31$

m $234 \times 70 = 16\,380$

n $610 \times 30 = 18\,300$

Example

Fact: $62 \times 3 = 186$

1 $6.2 \times 3 = 18.6$

2 $62 \times 30 = 1860$

3 $6.2 \times 0.3 = 1.86$

2 *Dividing by 10 is the same as multiplying by 0·1.*

Dividing by 100 is the same as multiplying by 0·01.

Dividing by 1000 is the same as multiplying by 0·001.

Example

$31 \div 10 = 0·1 \times 31 = 3·1$

$31 \div 100 = 0·01 \times 31 = 0·31$

$31 \div 1000 = 0·001 \times 31 = 0·031$

Divide each number by 10, 100 and 1000 by multiplying by 0·1, 0·01 and 0·001.

a 26   b 50

c 84   d 73

e 243   f 719

Remember

Align the columns.

Enter this key sequence into your calculator.

You need:

● calculator

Repeat the key sequence with several different 1-digit numbers.

Write what you notice.

Investigate what happens when you:

● start with a 6-digit number

● press the instead of the .

Write what you notice.

What if you enter the largest number of digits your calculator will display?

BEEP BEEP

Number equations

● **Write and use simple expressions in words and formulae**

Copy and complete.

Find the missing numbers.

1 **a** $6 \times \boxed{} = 48$

    b $27 + \boxed{} = 50$

    c $\boxed{} \times 7 = 49$

    d $\boxed{} - 56 = 44$

    e $2 \times \boxed{} + 10 = 28$

    f $36 \div 4 = \boxed{}$

2 **a** $\frac{1}{2} \times \boxed{} = 327$

    b $4 \times \boxed{} = 200$

    c $\boxed{}^2 = 144$

    d $\boxed{} - 129 = 11$

    e $56 \times \boxed{} = 168$

    f $95 \div \boxed{} = 19$

3 **a** $72 \div \boxed{} + 2 = 10$

    b $34 \times \boxed{} = 136$

    c $\frac{1}{2} \times \boxed{} = 68$

    d $57 \times \boxed{} = 5700$

    e $\boxed{} \div 10 = 930$

    f $\frac{1}{4} \times 600 = \boxed{}$

1 Balance these equations so both sides of the scales are equal.

| 6y   24 | 7 × 9   50 + y | 27p   54 | y − 64   36 |

| $\frac{120}{n}$ 30 | 8p + 4   68 | 96 − c   49 | 6 × 8   4n |

2 Find the solution to these equations.

    a $25p = 239 - 139$

    b $72 + n = 200 - 56$

    c $163 - y = 11^2$

    d $157 + 89 = 300 - n$

    e $36 \times 3 = 12y$

    f $42y = 200 - 32$

    g $20\% \times n = 11 \times 3$

    h $p + 49 = 7 \times 9$

    i $y - 144 = 12^2$

    j $50\% \times 126 = \frac{1}{4} \times y$

    k $760 \div 10 = 227 - p$

    l $84 \times 5 = (6 \times 7)y$

    m $25n = 50 \times 24$

    n $1001 - 635 = 61p$

    o $4t = 8^2 + 48$

3 Write an equation using the letter n to represent the unknown number for each of these statements.

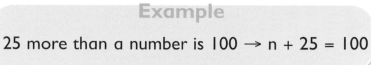

Example

25 more than a number is 100 → n + 25 = 100

   a 100 less than a number is 25

   b 60 times a number is 180

   c a number increased by 27 is 84

   d 44 times 4 is a number

   e a number decreased by 52 is 100

4 Find the value of n in the equations in question **3** .

1 How many different equations can you make using all of these cards?

2 Can you find the answer to each of your equations?

3 Add an extra card.

What equations can you make using all of the cards now?

Writing formulae

Follow the rule. Write the next 5 numbers in each sequence. Change the rule into a formula.

Example

Rule: Add 3 each time

17, 20, 23, 26, 29, 32, 35   Formula = n + 3

a The rule is: subtract 5 each time
→ 650, 645, ..., ...
  − 5
→ Formula =

b The rule is: add 25 each time
→ 375, 400, ..., ...
  + 25
→ Formula =

c The rule is: multiply by 3 each time
→ 2, 6 , ..., ...
  × 3
→ Formula =

d The rule is: divide by 2 each time
→ 256, ..., ...
  ÷ 2
→ Formula =

e The rule is: double each time
→ 30, ..., ...
  × 2
→ Formula =

I For each sequence:

- ● Identify the rule
- ● Write the next 5 numbers in each sequence
- ● Write the rule
- ● Devise a formula

Example

34, 38, 42, 46, 50, 54

If n is any number in the sequence, the rule is add 4 each time.   Formula = n + 4

a 3, 30, 300 . . .   The rule is . . .   Formula =

b 84, 90, 96 . . .   The rule is . . .   Formula =

c 3, 6, 12 . . .   The rule is . . .   Formula =

d 791, 781, 771 ...   The rule is ...   Formula =

e 2, 6, 18 ...   The rule is ...   Formula =

f 6, 13, 27...   The rule is ...   Formula =

g 8192, 4096, 2048 ...   The rule is ...   Formula =

h −250, −275, −300 ...   The rule is ...   Formula =

i 0·7, 0·9, 1·1 ...   The rule is ...   Formula =

2 The formula is  $T = n \times c$

T = total cost
n = number of items
c = cost per item

Apply the formula to find out the total cost of the number of items shown.
Copy and complete the tables.

a Cost = 12p

n	5	7	12	19	8	25	49	100
T								

b Cost = £22

n	6	9	20	15	11	26	32	50
T								

Write a rule and a number sequence for each formula below.

Choose a new start number each time for your sequence from the wheel.

a n + 5   b 2n − 1   c n × 10

d $\frac{1}{2}$n   e n − 15   f 2n + 2

g 3n   h 2n + n   i 5n − 1

Write 5 numbers in each sequence.

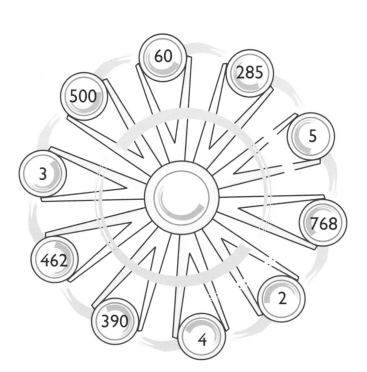

Divisibility testing time

 Write the following numbers.

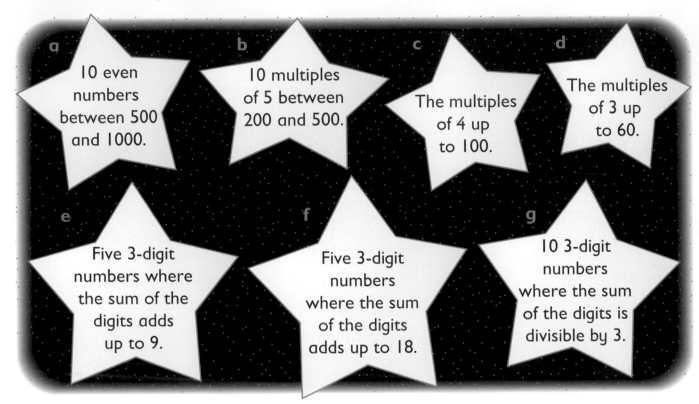

a 10 even numbers between 500 and 1000.

b 10 multiples of 5 between 200 and 500.

c The multiples of 4 up to 100.

d The multiples of 3 up to 60.

e Five 3-digit numbers where the sum of the digits adds up to 9.

f Five 3-digit numbers where the sum of the digits adds up to 18.

g 10 3-digit numbers where the sum of the digits is divisible by 3.

A number is divisible by:

2 if it is an even number and it ends in 0, 2, 4, 6 or 8

3 if the sum of its digits is divisible by 3

4 if the tens and units digits divide exactly by 4

5 if the last digit is 0 or 5

6 if it is even and it is also divisible by 3

8 if half of it is divisible by 4 or if its last three digits are divisible by 8

9 if the sum of its digits are divisible by 9

10 if the last digit is 0

 1 The boxes opposite have lost their labels. Use the divisibility tests to find which set of multiples are in each box and write a new label.

2 One number in each box does not belong. Write it down.

3 Explain how you worked out which numbers belong in each set.

a

4262		3750
	2756	
1364		4323
	3196	
3782		6148

Multiples of ...

b

4638		8397
	3420	
2393		3870
	2718	
7527		6804

Multiples of ...

c

4800		1872
	3120	
2688		3984
	1920	
3074		4652

Multiples of ...

d

3432		4792
	3520	
1388		2264
	4312	
1912		2176

Multiples of ...

e

1296		3132
	3528	
1512		1944
	2483	
2808		3060

Multiples of ...

f

4320		1601
	5610	
3243		2625
	2931	
7428		3735

Multiples of ...

4 Use the tests of divisibility to write 5 numbers:

a divisible by 3 between 400 and 500

b divisible by 4 between 600 and 2000

c divisible by 5 between 2000 and 2500

d divisible by 6 between 700 and 1500

e divisible by 9 between 4000 and 6000

f divisible by 8 between 200 and 800

Use your knowledge of divisibility tests to help you devise divisibility tests for other numbers.

Show examples of numbers that do not fit your test also, to show that your test works.

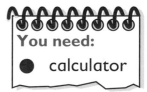

You need:
- calculator

- Write a list of multiples of 11.
- Devise a test of divisibility for 11.
- Try your test out on larger numbers. Use a calculator to check if your test works.

- Use the fact that 3 × 4 = 12 to develop a test for divisibilty by 12.
- Try your test out on larger numbers. Use a calculator to check if your test works.

Calculator round up

● **Use a calculator to solve problems**

It's round up time. Copy and complete each table.

Round answers up to the next whole number.

You need:

● calculator

a

Number of cattle	Divisor	Answer rounded up
49	14	
58	15	
67	17	
76	10	
85	15	
94	16	

b

Number of cattle	Divisor	Answer rounded up
101	20	
202	21	
303	24	
404	23	
505	22	
606	29	

1 Put these digits in the squares   **4   5   3**

☐ ☐ ÷ ☐ = ◯

to make an answer as close to the target as possible.

Round the quotients up each time.

a 2, 3, 6 → target 9

b 4, 6, 7 → target 8

You need:

● calculator

Example

3, 4, 5   → target 9

35 ÷ 4 = 8·75 → 9

2 Make 2 targets with these numbers.

a   4, 5, 8 → targets 10 and 17

b   5, 6, 9 → targets 14 and 16

c   3, 5, 8 → targets 17 and 29

3

$(C \times 1.8) + 32$

$(F - 32) \times 0.56$

Example

Celsius (C) to Fahrenheit (F)	Fahrenheit (F) to Celsius (C)
$(30\,°C \times 1.8) + 32 = 86\,°F$	$(15\,°F - 32) \times 0.56 \approx -10\,°C$

Use the formula on each boot to change the temperature.

Round your answers to the nearest whole number when necessary.

a   36 °F   c   77 °F   e   33 °C

b   28 °C   d   19 °F   f   9 °C

Back at the ranch, young Manuel has this problem to solve with his calculator.

a   Find the answers to 23·4 ÷ 2, 23·4 ÷ 3, 23·4 ÷ 4 … up to 23·4 ÷ 9.
Write as many decimal places as your calculator shows.

b   Check the answers with the inverse operation.

c   Explain what happens with each check.

You need:
● calculator

Using square numbers

● **Know square numbers to 12 × 12 and the squares of multiples of 10**

 1 Each board shows a square of pegs but some are covered over.

Work out how many pegs are in each square.

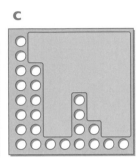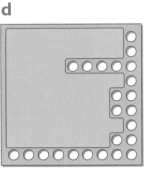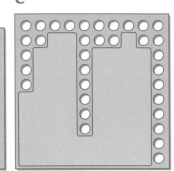

2 How many pegs will there be in

a a 12 × 12 square?   d a 60 × 60 square?

b a 20 × 20 square?   e a 80 × 80 square?

c a 30 × 30 square?   f a 50 × 50 square?

 1 Copy and complete.

a $4^2 + 2^2 =$ d $9^2 + 5^2 =$ g $12^2 + 8^2 =$

b $6^2 + 3^2 =$ e $10^2 + 8^2 =$ h $11^2 + 6^2 =$

c $8^2 + 7^2 =$ f $7^2 + 4^2 =$ i $6^2 + 9^2 =$

2 Copy and complete.

a $4^2 - 2^2 =$ d $10^2 - 5^2 =$ g $12^2 - 1^2 =$

b $5^2 - 3^2 =$ e $11^2 - 7^2 =$ h $11^2 - 9^2 =$

c $8^2 - 6^2 =$ f $6^2 - 4^2 =$ i $7^2 - 3^2 =$

3 **a** Copy and complete the column.

$2^2 - 1^2 =$
$3^2 - 2^2 =$
$4^2 - 3^2 =$
$5^2 - 4^2 =$
$6^2 - 5^2 =$
$7^2 - 6^2 =$

b Extend the column to $10^2 - 9^2$.

What pattern do you notice?

c Use the pattern to work out:

i $21^2 - 20^2 =$

ii $32^2 - 31^2 =$

iii $50^2 - 49^2 =$

iv $100^2 - 99^2 =$

4 What if the column of squares began like this?

$3^2 - 1^2 =$
$4^2 - 2^2 =$
$5^2 - 3^2 =$

Look for a pattern and extend the column to $10^2 - 8^2$.

More square number patterns

1 Using your calculator, find the squares of these numbers:

15, 25, 35, 45 and 55.

Record your results in a column.

2 Look for a pattern and use it to predict the answer to:

65^2, 75^2, 85^2 and 95^2.

Check your answers with the calculator.

3 Now extend to these square numbers:

105^2, 115^2, 125^2 and so on.

4 Write a rule for the square of numbers with units digit of 5.

You need:

● calculator

Example

$15^2 = 225$

$25^2 =$

$35^2 =$

Prime numbers

 Sort these numbers into 2 groups.

PRIME and COMPOSITE

Example

Prime: 2, 3 . . .   Composite: 4, 6 . . .

45 37 100 13 69 54 9 24 44 15 29 57 26 18 79 20 88 90 11 5

 1 Every number is the product of prime numbers, for example: 12 = 2 × 2 × 3

Use a factor tree to find which prime numbers form each product.

a 18   d 30   g 32

b 20   e 25   h 72

c 42   f 28   i 54

Example

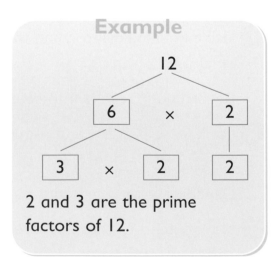

12

6 × 2

3 × 2 2

2 and 3 are the prime factors of 12.

2 Build 3 different factor trees for each of these numbers.

a 40   b 24   c 64

3 Build 4 different factor trees for each of these numbers.

a 48   b 100   c 36

4 Look closely at the factor trees you have made for the numbers in questions **2** and **3** .

a Explain why the factor trees are different for the same numbers.

b What is the same about the factor trees for the same numbers? Why?

5 Build factor trees for these numbers.

a 15   b 33   c 45   d 21

e What do you notice?

Prime numbers to 100 are easy to find!

1 How many prime numbers are there between 100 and 200?

Make a list.

2 Are there more prime numbers between 1 and 100 or 100 and 200?

3 Are they the same numbers with just 100 added on?

1	2	3	4	5	6	7	8	9	10
11	12	13	14	15	16	17	18	19	20
21	22	23	24	25	26	27	28	29	30
31	32	33	34	35	36	37	38	39	40
41	42	43	44	45	46	47	48	49	50
51	52	53	54	55	56	57	58	59	60
61	62	63	64	65	66	67	68	69	70
71	72	73	74	75	76	77	78	79	80
81	82	83	84	85	86	87	88	89	90
91	92	93	94	95	96	97	98	99	100

Tail-ends tables

● **Describe and explain sequences, patterns and relationships**

1 Refer to Grid 1, the multiplication table, and complete Grid 2 by writing the units digit of each multiple in the correct square. For example, for 14, write 4; for 28, write 8.

2 In Grid 2, colour these digits as follows:

0 red   5 blue

1 green   9 yellow

3 Write about the patterns you notice.

You need:

● RCM 19: Multiplication grids

● colouring pencils

Grid 2

×	1	2	3	4	5	6	7	8	9
1		2							
2									
3		6							
4									
5									
6									
7		4							

1 Refer to Grid 1, the multiplication table, and complete Grid 2 by writing the units digit of each multiple in the correct square. For example, for 14, write 4; for 28, write 8.

2 Look at Grid 2.

   a Which rows and columns have even digits only?

   b What patterns do you notice in columns 4 and 6?

You need:

● RCM 19: Multiplication grids

● ruler

Grid 2

×	1	2	3	4	5	6	7	8	9
1		2							
2									
3		6							
4									
5									
6									
7		4							
8									
9									

3 **a** In Grid 2, find all the digits 4 and 6.

Write them in the corresponding squares in Grid 3.

b In Grid 2, find all the digits 2 and 8.

Write them in their squares in Grid 4.

c In Grid 2, find all the digits 3 and 7.

Write them in their squares in Grid 5.

d In Grid 2, find all the digits 1 and 9.

Write them in their squares in Grid 6.

4 **a** In Grid 5, join all the 3s with straight lines to make a polygon.

Do the same for the 7s.

b In Grid 6, do the same for the digits 1 and 9.

5 For each Grid 3 to 6, write at least one statement about the patterns and symmetry.

Jan outlined a 2 x 2 square on Grid 1.

She multiplied the diagonally opposite numbers and found an interesting result.

She repeated this several times and found that the same thing happened.

Grid 1

×	1	2	3	4	5	6
1	1	2	3	4	5	
2	2	4	6	8	10	
3	3	6	9	12		
4	4	8				

You need:

● RCM 19: Multiplication grids

● ruler

1 **a** What did she discover?

b Will this be true for any 2 × 2 square on Grid 1? Investigate.

4	6
6	9

$4 \times 9 =$

$6 \times 6 =$

2 What if she outlined a 3 × 3 square and multiplied the corner numbers?

Will this work for a 4 × 4 square ... a 7 × 7 square?

Matchstick problems

 These three patterns are made with matchsticks.

 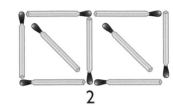 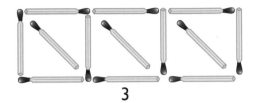

1 2   3

1 Draw the next two patterns.

2 Record the number of matches for the first five patterns in the table.

Pattern number	1	2	3	4	5	6	7
Number of matchsticks							

3 Find how many matches are added to make each new pattern.

4 Use this information to find the number of matches for the 6th and 7th pattern.

 These three patterns are made from matchsticks.

1 Use the pattern to complete the table.

Pattern number (P)	1	2	3	4	5	6	7
Number of matchsticks (M)	6						

2 Find how many matches are added to make each new pattern.

3 Write a formula in terms of P and M that is true for every example.

4 Test your formula by substitution for:

  a pattern 6   **b** pattern 10   **c** pattern 50

5 Copy and complete the table for this matchstick pattern.

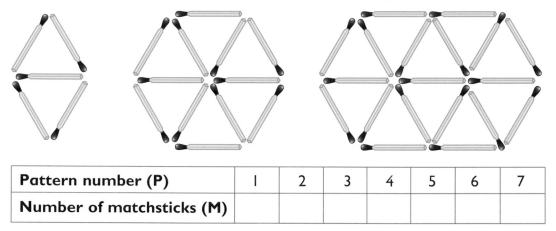

Pattern number (P)	1	2	3	4	5	6	7
Number of matchsticks (M)							

6 Find how many matches are added to make each new pattern.

7 Write a formula in terms of P and M that is true for every example.

8 Test your formula by substitution for:

   a pattern 6   **b** pattern 10   **c** pattern 50

1 Draw the next pattern in the sequence.

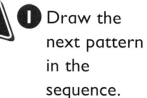

Pattern 1

Pattern 2

Pattern 3

2 Copy and complete the table to show the number of matchsticks used for each pattern.

Pattern number (P)	1	2	3	4	5	6	7
Number of matchsticks (M)							

3 Write a formula and use it to find the number of matchsticks needed for:

   a the 20th pattern   **b** the nth pattern

Mirror lines

Copy these shapes onto squared paper.

Draw the mirror line.

Using the mirror line, find and draw the reflected shape.

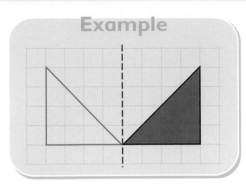

Example

You need:

- squared paper
- ruler
- mirror

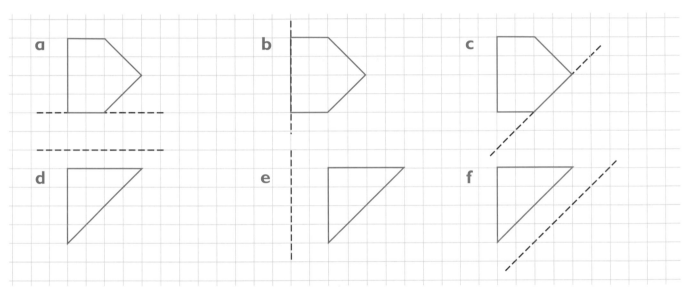

a

b

c

d

e

f

1 Copy each shape and mirror line onto squared paper.

Use the mirror line to find and draw the reflected shape.

You need:

- squared paper
- ruler

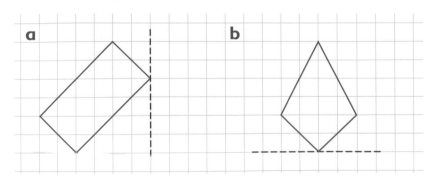

a

b

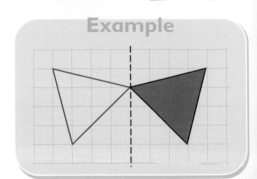

Example

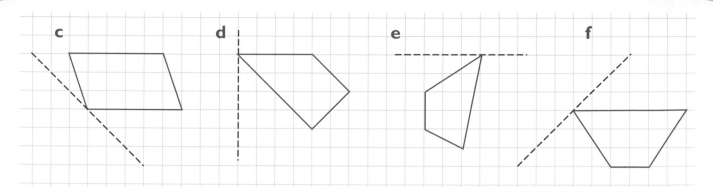

c   d   e   f

2 Draw accurately the reflections of these pentominoes.

Example

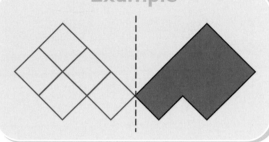

a

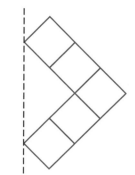

b

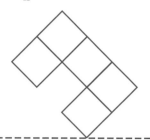

c

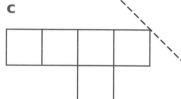

d

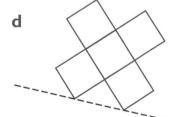

e

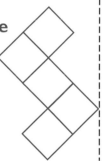

1 This is half a shape.

Draw 3 different whole shapes the original could have been.

Mark the lines of symmetry.

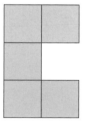

You need:

- squared paper
- a ruler

2 Do the same for these half shapes.

a

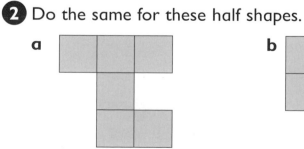

b

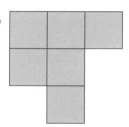

c

Four-way reflection

● **Recognise where a shape will be after reflection in two mirror lines at right angles**

 Copy these shapes on to your Resourse Copymaster. Use a different grid for each question. For each shape, the dotted line is the mirror line.

1 a Complete the rectangle ABCD.

b Write the co-ordinates of B and C.

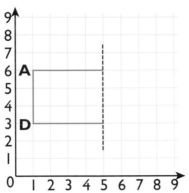

2 a Complete the square PQRS.

b Write the co-ordinates of Q and R.

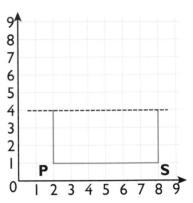

3 a Complete the rectangle EFGH.

b Write the co-ordinates of F and G.

4 a Complete the isosceles trapezium KLMN.

b Write the co-ordinates of M and N.

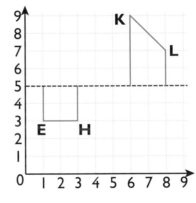

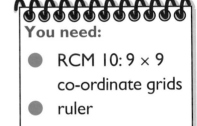

You need:

● RCM 10: 9 × 9 co-ordinate grids
● ruler

 Use a different grid on your Resource Copymaster for each question. In each grid, the vertical and horizontal mirror lines are drawn. Reflect each shape first in the vertical axis of symmetry then reflect both shapes in the horizontal axis of symmetry.

You need:

● RCM 12: 12 x 12 co-ordinate grids   ● ruler

Example

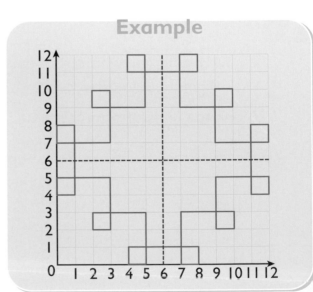

a
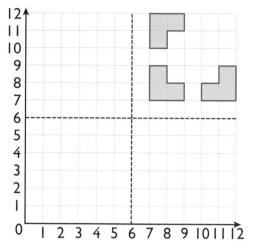

b
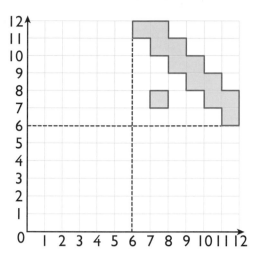

c
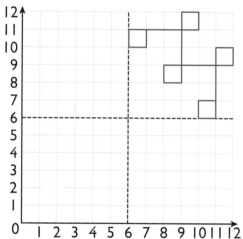

d
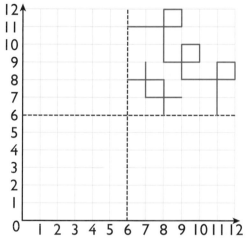

The mirror lines for this star shape are shown on the grid.

① Copy the shape on to one of the grids on your Resource Copymaster.

② Copy and complete this table to show the co-ordinates of points B, C and D reflected into all 4 quadrants.

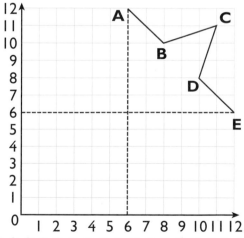

1st quadrant	2nd quadrant	3rd quadrant	4th quadrant
B = (8, 10)	B_1 =	B_2 =	B_3 =
C =	C_1 =	C_2 =	C_3 =
D =	D_1 =	D_2 =	D_3 =

You need:

● RCM 12: 12 x 12   ● ruler
   co-ordinate grids

Trapezium and kite

A trapezium has one pair of opposite parallel sides.

A kite has 2 pairs of adjacent sides equal.

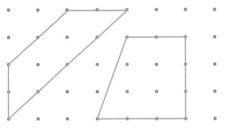

1 On 1 cm square dot paper draw:

   a 5 different trapezia

   b 5 different kites

2 Write the name trapezium or kite beneath each shape.

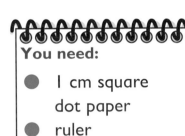

You need:

● 1 cm square dot paper

● ruler

1 Name each shape.

2 Write the letter of each shape which has:

   a only one pair of opposite parallel sides

   b 2 pairs of adjacent sides equal

   c only one line of symmetry

A   **B**   **C**   **D**

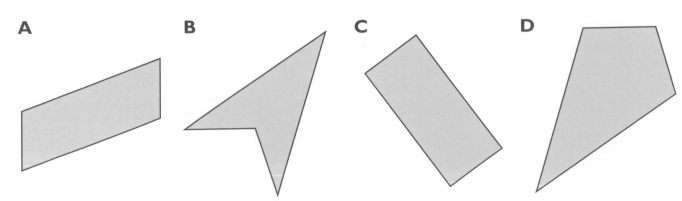

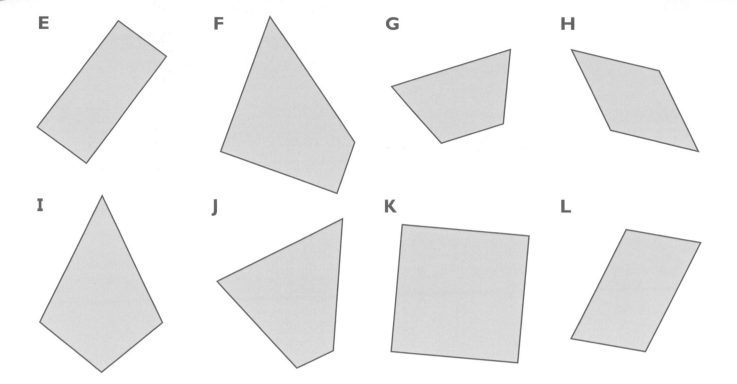

E   F   G   H

I   J   K   L

Jigsaw trapezia

1 Draw 4 congruent trapezia on triangular dot paper.

Cut out the shapes.

Assemble the 4 trapezia to make one large trapezium.

2 Draw and cut out 5 more trapezia.

Assemble all 9 trapezia to make one large trapezium.

3 Compare the large trapezia you made in questions **1** and **2**.

What can you say about their areas?

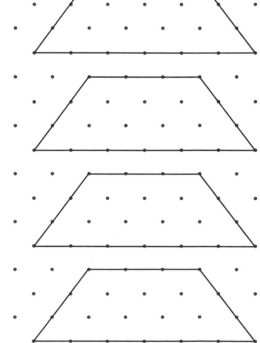

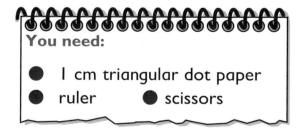

You need:

● 1 cm triangular dot paper
● ruler   ● scissors

Investigating diagonals

 ❶ Copy each quadrilateral on to 1 cm squared paper.

❷ Draw in the diagonals.

❸ Below each shape write its name.

quadrilateral names:
square, rectangle, rhombus, trapezium, parallelogram, kite

You need:
- 1 cm squared paper
- red pencil

❹ Draw in red the diagonals which intersect at right angles.

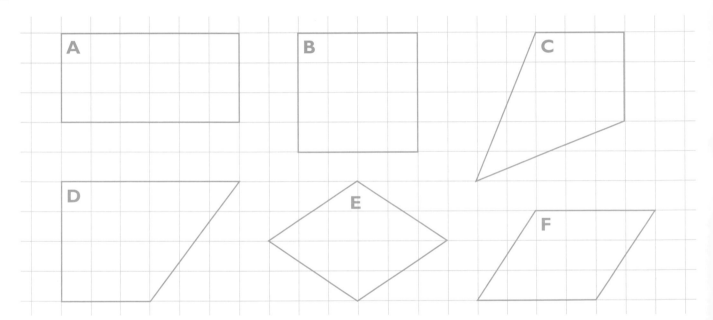

 ❶ Copy these quadrilaterals on to 1 cm squared paper.

❷ Draw in the diagonals and cut out each shape.

You need:
- 1 cm squared paper
- ruler ● scissors

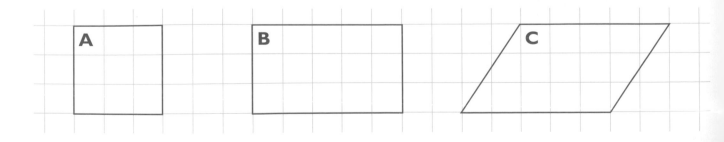

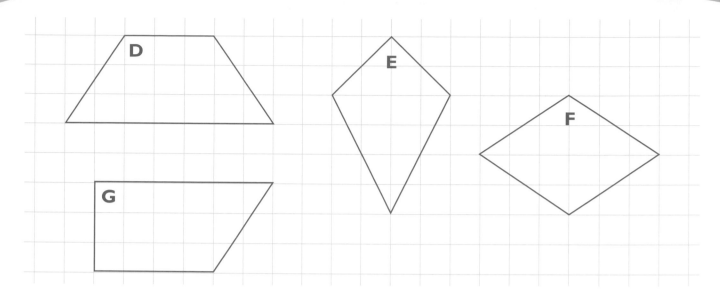

3 Copy and complete. Mark ✓ for *yes* and ✗ for *no*.
Find the answers by folding and measuring.

Property	Quadrilateral						
	A	B	C	D	E	F	G
4 sides equal	✓	✗	✗				
4 angles equal	✓						
diagonals are same length							
diagonals cut each other in half							
diagonals intersect at right angles							
a diagonal is an axis of symmetry							

Diagonals of rectangles and parallelograms.

1 Investigate this statement.
The diagonals of any rectangle are equal and bisect each other.

2 Draw about 6 different parallelograms on 1 cm square dot paper.
Draw in the diagonals.
For each parallelogram, measure the diagonals from the vertices to the intersection.
Write what you notice.

3 Compare your results for questions **1** and **2**.
In what ways are the diagonals of rectangles and parallelograms similar? Different?

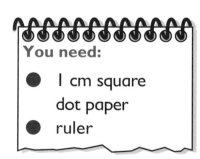

You need:
● 1 cm square dot paper
● ruler

Classifying quadrilaterals and triangles

● **Classify quadrilaterals and triangles**

1 Copy these quadrilaterals on to the 3 × 3 pinboards.

2 For each quadrilateral:

a write its name

b mark the equal sides

c mark each pair of equal angles in the same colour

d draw the axes of symmetry.

You need:

● RCM 2: 3 × 3 pinboards
● ruler
● colouring materials

a
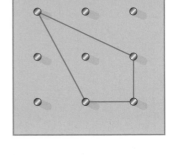

b

c
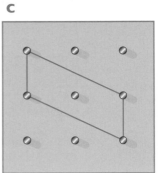

d

1 On RCM 2, draw 16 different quadrilaterals. Include those drawn in the ▢ activity.

Example

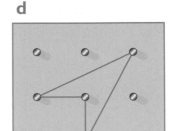

You need:

● RCM 2: 3 × 3 pinboards
● ruler
● colouring materials

2 Name each quadrilateral.

3 Mark the parallel sides >>.

4 Mark the perpendicular sides ⌐ .

5 Mark each pair of equal angles in the same colour.

6 Copy and complete the table.

For each property, record the number for these shapes: parallelogram, kite and trapezium.

Property	parallelogram	kite	trapezium
opposite angles equal			
one or two right angles			1
one pair of parallel sides			1
two pairs of parallel sides			
perpendicular sides			1

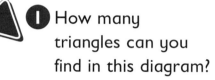

1 How many triangles can you find in this diagram?

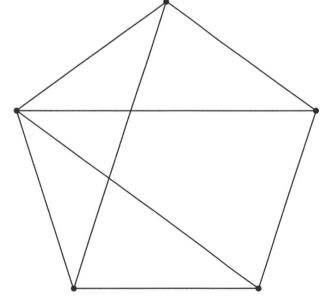

2 How many triangles are:

   a isosceles?

   b scalene?

3 Draw a regular pentagon.

Then draw in all the diagonals.

How many triangles of different sizes can you find?

Draw a different pentagon for each triangle and colour the triangle.

You need:

● 2-D shape: pentagon
● ruler
● colouring materials

Whale weights

 1 Convert these weights to decimal notation.

a $\frac{3}{10}$ kg, $\frac{3}{100}$ kg, $\frac{3}{1000}$ kg

b $\frac{6}{10}$ kg, $\frac{6}{100}$ kg, $\frac{6}{1000}$ kg

2 Convert these weights to fractional notation.

a 0·9 kg,   0·09 kg,   0·009 kg

b 0·7 kg,   0·07 kg,   0·007 kg

3 Copy and complete.

plastic pen caps	weight (fraction)	weight (decimal)
1000	1 kg	1·0 kg
100	$\frac{1}{10}$ kg	
		0·01 kg
1	$\frac{1}{1000}$ kg	

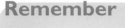

Remember

$\frac{1}{10}$ kg = 0·1 kg

$\frac{1}{100}$ kg = 0·01 kg

$\frac{1}{1000}$ kg = 0·001 kg

1 Copy and complete this table. Round each of these weights to the nearest tenth of a kilogram and the nearest kilogram.

weight	rounded to nearest $\frac{1}{10}$ kg	rounded to nearest kg
83·635 kg	83·6 kg	84 kg
46·270 kg		
67·475 kg		
59·520 kg		
70·090 kg		
112·875 kg		
111·040 kg		
94·760 kg		

2 Copy and complete this table which shows the average weight of adult whales.

whale	tonnes	kilograms
blue whale	150 t	150 000 kg
sei whale	20 t	
humpback whale	25 t	
killer whale (male)		9000 kg
killer whale (female)		6400 kg
narwhal	1·5 t	
minke whale	8 t	
sperm whale		57 000 kg
pygmy sperm whale		400 kg
dwarf sperm whale		210 kg

3 Find how many times heavier the blue whale is than:

a the humpback whale

b the narwhal

c the pygmy sperm whale

1 Calculate, in kilograms:

a The weight of krill a blue whale will eat in one day.

b The difference between the average and heaviest recorded weights of a blue whale.

2 A slice of pizza for a human weighs 0·2 kg. How many times larger would the pizza slice have to be for the appetite of an average blue whale?

3 What is the approximate weight of a one-month old baby blue whale?

Facts about blue whales

● The largest living creatures on Earth.

● The average weight is about 150 tonnes.

● The heaviest ever caught weighed about 190 tonnes.

● The daily consumption of krill, small sea creatures similar to shrimp, is about 4 tonnes.

● Birth weight is about 3000 kg.

● A baby blue whale puts on about 100 kg in weight each day.

Pounds and ounces

● **Read and answer questions about scales involving imperial units (mass)**

 1 Copy and complete.

a 1 pound = 16 ounces

   1 quarter pound = ☐ ounces

   1 half pound = ☐ ounces

   3 quarters of a pound = ☐ ounces

b 1 lb = 16 oz

   $\frac{1}{4}$ lb = ☐ oz

   $\frac{1}{2}$ lb = ☐ oz

   $\frac{3}{4}$ lb = ☐ oz

2 Write **true** or **false** for each statement.
Use the conversion scale on the opposite page to check each answer.

a 1 lb is about 450 g

b $\frac{1}{2}$ lb is more than 200 g

c 4 oz rounds up to 100 g

d 12 oz rounds down to 300 g

e 8 oz > 200 g

f $\frac{1}{4}$ lb < 100 g

g 2 lb is just over 900 g

h 1000 g < 16 oz

> **Example**
>
> 2 lb is less than
> 1000 g. True

1 Copy and complete.

1 lb ≈ 454 g $\frac{1}{4}$ lb ≈ ☐

$\frac{1}{2}$ lb ≈ ☐ $\frac{3}{4}$ lb ≈ ☐

2 Use the conversion graph on the right
to work out these answers.

Round the figures to the nearest
ounce or gram.

a 7 oz ≈ ☐ g

b ☐ oz ≈ 30 g

c 14 oz ≈ ☐ g

d 2 lb ≈ ☐ g

e ☐ oz ≈ 250 g

f ☐ oz ≈ 370 g

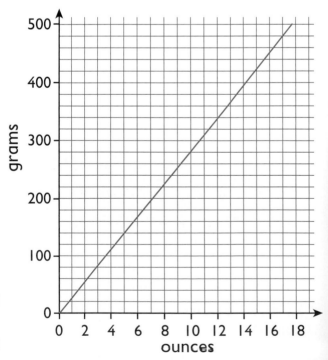

3 Copy and complete the table.

Item	Metric	Imperial
cornflakes	500 g	
marmalade	340 g	
butter		7 oz
tin of tuna	90 g	
tea		$\frac{1}{4}$ lb
bread		$1\frac{3}{4}$ lb

Remember

2·2 lbs is approximately 1 kilogram.

4 Kate wants to knit a sweater for her baby brother.

The old knitting pattern which she got from her gran needs five 1 oz balls of wool.

If knitting wool is sold in balls of 50 g, how many balls of wool will she need to buy?

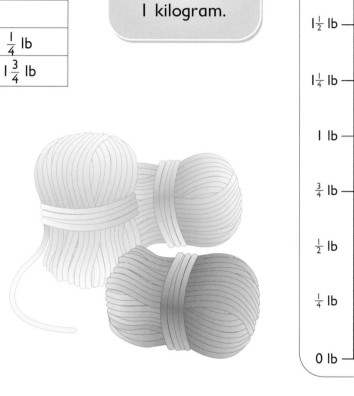

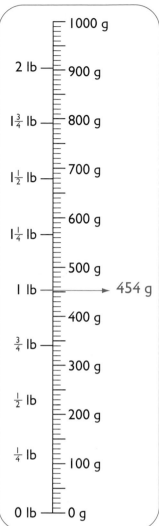

1000 g
2 lb — 900 g
$1\frac{3}{4}$ lb — 800 g
$1\frac{1}{2}$ lb — 700 g
600 g
$1\frac{1}{4}$ lb —
500 g
1 lb — → 454 g
400 g
$\frac{3}{4}$ lb —
300 g
$\frac{1}{2}$ lb —
200 g
$\frac{1}{4}$ lb —
100 g
0 lb — 0 g

The imperial system of weights used this series of numbers:

1, 2, 4, 8, 16, 32 .... and so on.

Write the least number of weights you need to balance an object weighing:

a 11 oz   **f** 47 oz

b 15 oz   **g** 55 oz

c 23 oz   **h** 62 oz

d 30 oz   **i** 70 oz

e 43 oz   **j** 99 oz

Example

weight of apples = (1 + 4 + 8 + 32) oz
= 45 oz

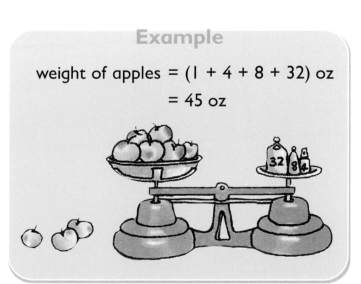

Mass in action

- Convert measures between units and read and answer questions including decimals
- Read and answer questions about scales involving imperial units (mass)

 1 Megan weighed 5 parcels labelled A to F.

Write the weight of each parcel in grams as shown by the arrows on the scale.

2 Now write each weight in kilograms.

3 Find the two parcels which together will weigh the same as parcel F.

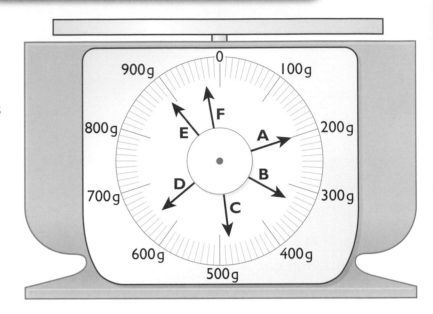

 1 Calculate the total weight of each pair of parcels then convert this weight to pounds.

a

3·4 kg   4·6 kg

b

6·1 kg   5·9 kg

You need:
- calculator

c

4·7 kg   12·3 kg

d

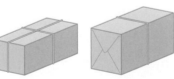

17·5 kg   14·5 kg

Example

1 kg ≈ 2·2 lb

8 kg ≈ 8 × 2·2 lb

≈ 17·6 lb

e

18·8 kg   15·2 kg

f

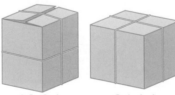

23·9 kg   26·1 kg

2 Find the weight of each sports bag in kilograms.

Round your answers to one decimal place.

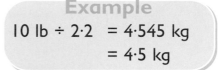

Example

10 lb ÷ 2·2 = 4·545 kg

= 4·5 kg

a

15 lb

b

20 lb

c

24 lb

d

38 lb

e

33 lb

f

19 lb

3 The scales show the amount of fat in 85 g servings of food.

Find which food has approximately:

a 10% fat   **b** 20% fat

c 25% fat   **d** 7% fat

fried cod

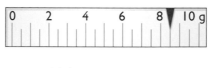

beefburger

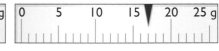

roast chicken

sausage

Household rubbish

A recent survey found that an adult produces 400 kg of household rubbish per year. This table shows how it is made up.

1 Use your calculator to work out the weight of rubbish in each category.

2 A household produces 1·2 t of rubbish in a year. How many adults live in the house?

3 a How much rubbish would an adult produce in a year if he recycled the glass, paper and card?

b What if he also had a compost heap at the foot of his garden for his vegetable and decomposable material?

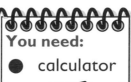

You need:
● calculator

Category of rubbish

33%	paper and card
20%	vegetable, decomposable
11%	plastics
9%	glass
8%	dust, cinders etc.
7%	metal
2%	textiles
10%	miscellaneous

Mostly medians

● **Find the median of a set of data**

 1 Find the median (middle number).

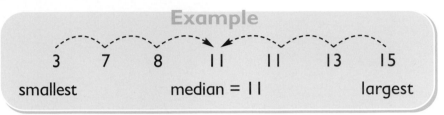

Example

3 7 8 **11** 11 13 15

smallest   median = 11   largest

a 2, 5, 9

b 1, 1, 2, 3, 7

c 4, 6, 6, 7, 8

d 2, 2, 2, 6, 6, 6, 6

e 50, 60, 80, 80, 90

f 19, 28, 51

2 Find the median (halfway between the middle two numbers).

Example

20   20   26   30

Median = 23

Middle two numbers are 20 and 26.
To find halfway between them:

20 + 26 = 46

46 ÷ 2 = 23

a 5, 7, 9, 10

b 2, 2, 3, 4, 6, 8

c 0, 0, 0, 4, 4, 5

d 10, 10, 20, 20, 30, 40, 60, 70

e 1, 1, 2, 3, 3, 4, 5, 5, 6, 8

f 52, 53, 53, 56

3 Find the median of these numbers.

a 4, 9, 10

b 1, 1, 5, 6

c 2, 2, 4, 9, 10

d 20, 50, 50, 55, 60, 90

e 0, 0, 5, 9, 14, 23, 27

f 7, 7, 8, 8, 8, 9, 9, 9, 10, 10

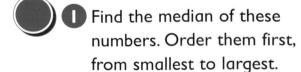 **1** Find the median of these numbers. Order them first, from smallest to largest.

Example

3 7 8 11 11 13 15

smallest   median = 11   largest

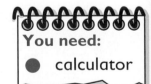

You need:
● calculator

a 5, 9, 3, 6, 4

b 2, 8, 2, 8, 2, 8, 2

c 4, 4, 9, 9, 4, 4, 9, 9, 4

d 50, 30, 10, 30, 20

e 18, 34, 20, 28, 69, 42, 38

f 2, 6, 3, 3, 2, 1, 7, 4, 9, 1, 4

2 Find the median of these values. Order them first, then find the middle two numbers. The median is halfway between them.

a 9, 3, 2, 5

b 6, 2, 2, 6, 2, 6

c 12, 2, 5, 10, 14, 8

d 5, 4, 4, 4

e 60, 10, 50, 30

f 700, 200, 300, 100, 500, 400

3 Find the median of these values.

a 249, 583, 729, 419, 700

b 3000, 600, 5000, 900, 2400, 750

c 964, 835, 694

d 2420, 2950

e 3200, 5800, 9200, 4600, 7200, 3800, 4900

f 1023, 3051, 2709, 5308, 9025, 4352, 6951, 3502

4 Find the median of these quantities.

a 30 g, 64 g, 20 g, 80 g, 240 g

b £4, £2.50, £12, £16.50, £11, £5.50, £9

c 140 cm, 250 cm, 160 cm, 300 cm, 92 cm, 200 cm

d £1.29, 75p, £2.43, £1, 98p

e 7 g, 2 g, 2 g, 9 g, 5 g, 5 g, 11 g, 2 g, 6 g, 8 g, 11 g, 8 g

f 2500 ml, 600 ml, 1600 ml, 400 ml

Sandy counted the crisps in the packets of different brands.

1 Calculate the mode for each brand. Which brand has the highest mode?

31, 35, 39, 30, 40, 38, 36, 30, 37

35, 37, 34, 35, 36, 34, 35, 37

39, 22, 48, 45, 30, 24, 27, 32, 23, 27

2 Calculate the median for each brand. Which brand has the highest median?

3 Calculate the range for each brand. Which brand has the narrowest range?

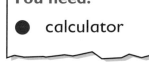

You need:

● calculator

4 If each brand costs the same, which is the best value? Explain your answer.

Mostly means

● Find the mean of a set of data

1 Find the mean of each pair of numbers.

Example

10, 22   Total = 10 + 22 = 32   32 ÷ 2 = 16   Mean = 16

a 6, 20
b 12, 30

c 60, 90
d 100, 500

e 72, 24
f 230, 180

2 Find the mean of these numbers.

Example

4, 8, 10, 14   Total = 4 + 8 + 10 + 14 = 36   36 ÷ 4 = 9   Mean = 9

a 6, 9, 12
b 4, 5, 9

c 1, 1, 10
d 7, 7, 7

e 100, 20, 30
f 0, 8, 10

3 Find the mean of these numbers.

a 5, 5, 7, 7
b 2, 0, 7, 7

c 10, 10, 10, 50
d 50, 100, 100, 150

e 6, 12, 6, 12
f 4, 19, 7, 2

4 Find the mean of these numbers.

a 1, 4, 1, 4, 5
b 5, 20, 30, 40, 5

c 0, 0, 10, 20, 30
d 10, 40, 50, 50, 100

e 9, 0, 1, 0, 5
f 500, 100, 200, 100, 100

1 Find the mean of these numbers.

a 4, 14
b 5, 5, 20
c 3, 3, 7, 7
d 0, 0, 3, 4, 5

e 10, 10, 10, 20, 20
f 1, 2, 3, 1, 2, 3, 1, 3
g 1, 4, 1, 4, 1, 4, 6
h 10, 10, 5, 10, 15, 10, 5, 5, 10, 10

You need:

● calculator

2 Find the mean of these weights.

a 100 g, 400 g
b 1 g, 2 g, 2 g, 5 g, 10 g
c 100 g, 300 g, 100 g, 300 g
d 10 g, 20 g, 20 g, 40 g, 50 g, 100 g
e 1 g, 10 g, 5 g, 2 g, 25 g, 2 g, 5 g, 10 g, 21 g
f 20 g, 45 g, 30 g, 10 g, 15 g, 25 g, 20 g, 15 g, 20 g, 30 g

3 Find the mean of these numbers. Use your calculator to help.

a 94, 73

b 48, 60, 72, 38

c 231, 162, 453

d 7, 17, 27, 37, 47, 57

e 1250, 4890, 2720, 1990

f 1, 101, 110, 11, 101, 111, 1, 110

g 9, 17, 28, 39, 46, 58, 71, 93, 104, 129

h 8, 6, 2, 5, 9, 1, 7

4 Find the mean of these prices. Use your calculator to help.

a £7, £4

b £9, £12, £6, £10

c £67, £93, £84, £32, £99, £18

d £2.50, £1.75 £3.01

e £1.20, £7.40, £3.10, £9.20, £6.60, £2, £4.80, £9·70

f £2, £5, £3, £1, £7, £8.50, £2.50, £1, £1, £10

g 30p, £1.80, 60p, £3.90, 80p

h 73p, £1.80, 97p, 82p, £2.30, £1.78

5 Calculate the mean weight of a sweet.

TOTAL WEIGHT = 45g

TOTAL WEIGHT = 82g

TOTAL WEIGHT = 72g

TOTAL WEIGHT = 200g

a 5 sweets

b 10 sweets

c 16 sweets

d 25 sweets

Find the median, mean and range of these values.

HINT
There may be two middle values.

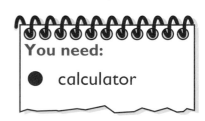
You need:
● calculator

a 25, 91, 63, 84, 32

b 324 g, 596 g, 280 g, 300 g, 146 g, 63 g, 92 g, 451 g

c £1.25, £4.62, £5.88, £9.27, £16.41, £8.47

d 31·2 cm, 51·1 cm, 18·5 cm

e 400, 9000, 3500, 250, 850, 1250, 500, 7000, 6250, 4950

f 126 kg, 621 kg, 126 kg, 162 kg, 261 kg, 216 kg, 609 kg

g 319 ml, 913 ml, 139 ml, 319 ml

h 16 seconds, 94 seconds, 31 seconds, 75 seconds, 13 seconds

i 94p, £1.84, £1, 34p, 72p, £1.96, 50p, £1, £2.14, £5.16

j 3 m, 214 cm, 1·6 m, 91 cm, 2 m, 250 cm

Average performance

● Solve problems using the mode, range, median and mean

 Here are the results of 10 hockey matches:

2-0 1-1 3-0 0-0 4-2 3-1 2-2 6-2 1-3 2-1

1 For each match, find the total number of goals scored.

2 Write the totals in order, from smallest to largest.

3 Find the

    a mode   **c** mean

    b median   **d** range

 1 Find the mode, median, mean and range for each of the following football teams.

You need:

● calculator

Everham	Sandstead United	Witherley
2	3	1
0	2	0
1	5	3
2	0	1
4	1	0
2	2	0
8	3	4
0	2	1
1	9	2
3	0	2
3	2	2
1	1	0
5	3	3
0	2	1
1	10	0
2	3	1
4		2
1		0
0		1
3		2

2 Use your averages to compare the performances of the three football teams.

3 Which team was the most consistent? Explain your answer.

4 If each team scored 8 goals in the next match, which statistics would be affected?

1 Work with a partner. Use the sports section from the newspaper to find the results of a football team. You need about 20 results.

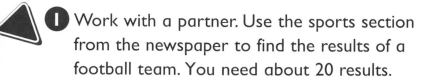

You need:

● the sports section from a newspaper
● calculator

2 Use statistics to compare the goals scored by the team with the goals scored by their opponents.

3 Find the results of another team. Use statistics to compare the goals they scored with the goals scored by the team in question **1**.

Weather watch

- Find the median, mean and range
- Represent data in different ways and understand its meaning

The graph shows how full Ardingly Reservoir was during 1996.
Each point shows the level at the beginning of the month.

1 During which months was the reservoir full?

2 During which month was the reservoir at the lowest level?

3 Estimate the level at the beginning of August.

4 When was the reservoir half full?

5 When was the reservoir 80% full?

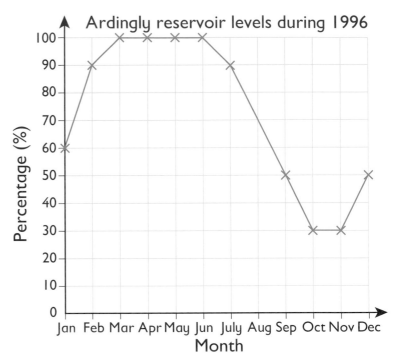

The table shows the outdoor temperatures at Pennington County Infant School on 6th July 2006.

You need:
- graph paper
- ruler

Pennington County Infant School (outdoor temperatures for 6th July 2006)	
Time	Temp (°C)
8 a.m.	16·7
10 a.m.	16·8
12 noon	19·5
2 p.m.	
4 p.m.	19·6
6 p.m.	16·2
8 p.m.	16·6
10 p.m.	16·9

1 Copy and complete the line graph.

2 Estimate the temperature at:

   a 2 p.m.

   b 11 a.m.

   c 5 p.m.

3 Estimate when the temperature was 17·5 °C.

4 Use temperatures in the table and your estimate for 2 p.m. to calculate the mean temperature from 8 a.m. to 10 p.m.

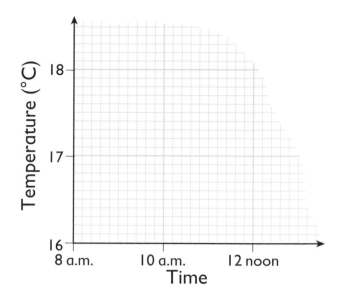

The graph shows the water temperatures of the Tioga River, Pennsylvania, USA for a week in 2004.

1 a Find the maximum temperature for each day.

   b Calculate the average maximum temperature for the week.

2 Find the average minimum temperature for the week.

3 a Calculate the change in temperature for each day.

   b Calculate the average daily change in temperature.

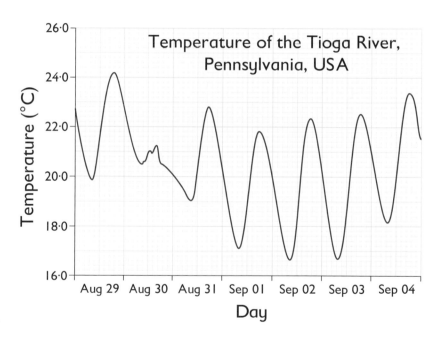

Spinners

Describe the chance of spinning each colour. Use the words on the right.

a

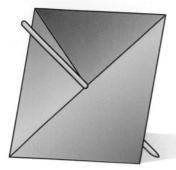

Red

Blue

Green

Orange

b

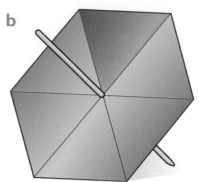

Red

Green

Red, green or blue

no chance

even chance

impossible

unlikely

likely

small chance

certain

good chance

1 Which spinners give:

  a each colour an equal chance

  b yellow no chance

  c blue and green, an equal chance

  d yellow, the smallest chance?

2 Which colour has the greatest chance and for which spinner?

3 Which colours has an even chance, and for which spinner?

spinner 1

spinner 2

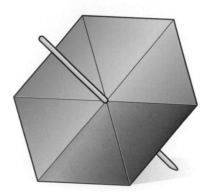

spinner 3

4 For this spinner, describe the chance of getting:

a   an even number

b   1

c   an odd number

d   a number less than 5

e   4

Mark these chances on a scale from impossible to certain.

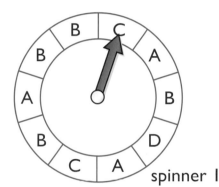

5 Which spinner gives the best chance of spinning a letter C?
Explain your answer.

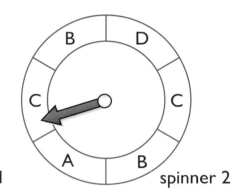

spinner 1   spinner 2

1 Cut out an 8-sided spinner.

Number it using the numbers 1, 2, 3, 4.
Their chances must not all be the same.

2 Predict the most likely numbers.

3 Describe the probability of each number happening.

4 Spin the spinner 30 times. Record the results in a tally chart.

5 Draw a bar line chart for the data.

6 Comment on your results.

You need:

● a copy of RCM 6: Spinners
● squared paper
● scissors   ● ruler
● paperclip (for spinner)

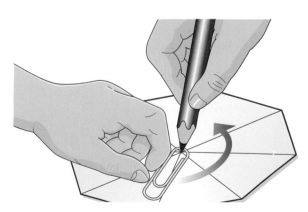

Spending pie charts

The pie charts show how Rachel and Peter spent their book vouchers.

Find the fraction they spent on each book.

Calculate the amount they spent on each book.

Copy and complete the tables.

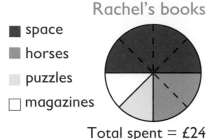

Rachel's books

■ space
■ horses
□ puzzles
□ magazines

Total spent = £24

Book	Fraction	Money spent (£)
space		
horses		
puzzles		
magazines	$\frac{1}{8}$	$\frac{1}{8}$ of 24 = 3
Total		

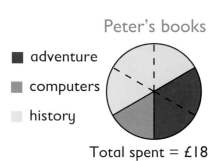

Peter's books

■ adventure
■ computers
□ history

Total spent = £18

Book	Fraction	Money spent (£)
adventure		
computers		
history		
Total		

You need:
● calculator

1 The pie chart show how Maria spent her birthday money.

Find the fraction she spent on each item.

Calculate the amount she spent on each item.

Copy and complete the table.

Maria

■ books ■ clothes
□ handbags

Total spent = £30

Item	Fraction	Money spent (£)
books		
Total		

2 The pie chart show how Jamaine spent her birthday money.

What fraction did she spend on magazines?

What was the total amount she spent?

Copy and complete the table.

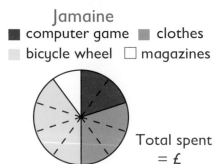

Jamaine

■ computer game ■ clothes
□ bicycle wheel □ magazines

Total spent = £

Item	Fraction	Money spent (£)
magazines		7
Total		

3 The pie charts show how two families spent their holiday money.

    a Who spent the most on entertainment? Explain your answer.

    b Make a table for each family.

Wilkins family

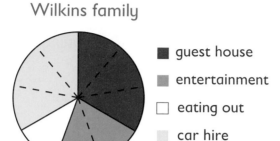

■ guest house

■ entertainment

□ eating out

▨ car hire

Total spend = £819

Baxter family

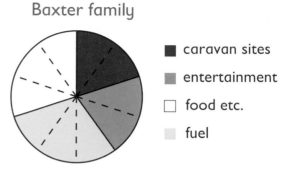

■ caravan sites

■ entertainment

□ food etc.

▨ fuel

Total spend = £740

1 The table shows how Solomon spent his pocket money.

    a Calculate the fraction he spent on each item.

    b Sketch a pie chart for the table.
Colour the sectors and make a key.

You need:

● a pair of compasses

● ruler   ● coloured pencils

Item	Money spent (£)	Fraction
Geometry set	2	
Sweets	1	
Cinema	3	
Magazines	2	
Total		

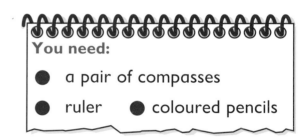

2 Draw pie charts for these tables.

a

Item	Money spent (£)
Calculator	8
Pencil set	4
Music CD	12
Total	

b

Item	Money spent (£)
Shoes	21
Dress	14
Trousers	21
Total	

Body statistics

● **Solve problems by collecting, processing and presenting data**

Marks says

People with long arms have long fingers.

You need:

● RCM 17: Planning an investigation

● pencil and paper

● tape measure

 1 Work with a partner. One person measures the length of the middle finger (right hand) for each child with a yellow badge (the children with longer arms).

The other person measures the length of the middle finger (right hand) for each child with a red badge (the children with shorter arms).

Make sure you don't measure the same person twice.

2 Find the mode of the finger size for the children with the longer arms and for the children with the shorter arms.

3 Do you think Mark is correct? Explain your answer.

 Work as a group. Investigate one of the following statements.

People with longer arms have bigger heads.

People with shorter arms have smaller feet.

People with shorter arms have narrower wrists.

1 Write down your individual prediction.

2 Decide how the group will collect and record each set of data from the children in your class.

3 Calculate statistics for each set of data.

4 Compare the statistics.

5 Does the data support your prediction? Write a sentence.

6 How could the investigation be improved? Write a sentence.

You need:

● RCM 17: Planning an investigation

● pencil and paper

● tape measure

1 Work with a partner or in a group. Think up your own investigation

For example

Children with small hands have small feet.

You need:

● RCM 17: Planning an investigation

● pencil and paper

● tape measure

2 Write down your individual prediction.

3 Use the information on the board to separate children with smaller and larger feet.

4 Plan and carry out your investigation.

5 Does the data support your prediction? Write a sentence.

6 How could the investigation be improved? Write a sentence.

Backpack weights

- **Convert smaller units to larger units (e.g. g to kg) and vice versa**
- **Add, subtract, multiply and divide whole numbers and decimals**

1 Look at the weight of each backpack in the ● section below.

Write each weight as:

a kilograms and grams

b grams only

Example

Delroy's pack:

5·248 kg = 5 kg 248 g

= 5248 g

2 Delroy used these weights to weigh his backpack.

In the same way, work out which weights were used to weigh the other 5 backpacks.

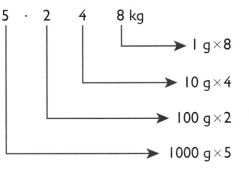

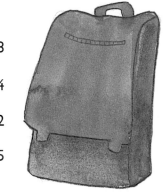

Six children are going on a school trip. The weight of each backpack is given below.

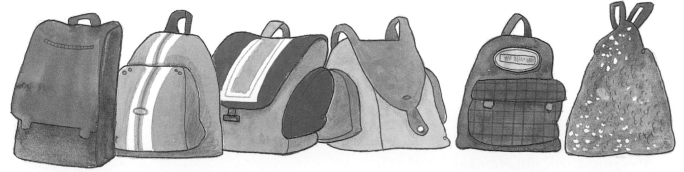

Delroy	Meera	Tom	Jenny	Sam	Amy
5·248 kg	5·813 kg	6·478 kg	4·685 kg	8·157 kg	7·589 kg

1 Write the number of grams represented by the 8 digit in the backpacks of Meera, Tom, Jenny and Sam.

2 Round each backpack weight to the nearest tenth of a kilogram and to the nearest kilogram. Record your results in a table.

Backpack	Rounded to nearest $\frac{1}{10}$ kg	Rounded to nearest kg
Delroy	5·2 kg	5 kg
Meera		

3 Find the difference in weight, in grams, between these backpacks:

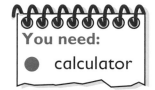

You need:
- calculator

   a Delroy's and Tom's   **b** Meera's and Sam's

   c Jenny's and Amy's   **d** the lightest and the heaviest

4 Find the cost of sending these packets by First Class Post.

   a 15 packets each weighing 330 g

   b 24 packets each weighing 650 g

   c 8 packets each weighing 190 g

   d 20 packets each weighing 960 g

Weight of packet	First Class Post
0 – 100 g	£1.00
101 – 250 g	£1.27
251 – 500 g	£1.70
501 – 750 g	£2.20
751 – 1000 g	£2.70

5 The total cost of posting some 220 g packets was £15.24. How many packets were posted?

1 Find which three backpacks will weigh the same as the combined weight of Sam's and Amy's pack.

2 Jenny and Amy are twins. If they decide to rearrange their bags so that each will carry the same weight, how many grams will Amy give to Jenny to carry?

Weights workout

● **Solve problems with several steps and decide how to carry out the calculation**

1 The table shows the number of copies of each newspaper ordered by two newsagents.

Using a calculator, work out the weight, in kilograms, of newspapers delivered to each newsagent's shop.

Copy and complete the table.

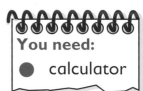

You need:
● calculator

Newspaper	Weight	Newsagent 1		Newsagent 2	
		Copies	Weight	Copies	Weight
Daily Times	270 g	20	5·4 kg	30	kg
Herald	300 g	50	kg	75	kg
Morning Express	350 g	30	kg	50	kg
Daily News	250 g	60	kg	40	kg
		Total weight	kg	Total weight	kg

2 Find the difference in the total weight of newspapers ordered by the two shops.

The table gives the weight of each daily and Sunday newspaper.

Newspaper	Daily Times	Herald	Morning Express	Daily News	Post on Sunday	Sunday News
Weight	270 g	300 g	350 g	250 g	400 g	500 g

You need:
● calculator

1 Calculate the weight, in kilograms, of newspapers delivered to each house in one week.

Address	Daily (Mon–Sat)	Sunday	Total in kg
4 Thorn Road	Morning Express	Sunday News	2·1kg + 0·5kg = 2·6kg
7 Briar Avenue	Herald	Post on Sunday	
35 Ash Grove	Daily Times	Sunday News	
60 Oak Lane	Daily News, Herald	Sunday News	
48 Elm Crescent	Morning Express, Daily Times	Post on Sunday	
16 Willow Way	Herald, Daily News	Post on Sunday	

2 At the start of Pat's paper round, his bag weighs 7·35 kg. He has 5 copies of the Daily Times, 6 of the Herald, 7 of the Morning Express and the rest are the Daily News.

How many copies of the Daily News does he deliver?

3 Peter delivers 17 copies of Post on Sunday and 13 copies of Sunday News to 20 houses.

a How many customers take two Sunday newspapers?

b How heavy is his bag at the start of his Sunday paper round?

1 Kenny, Liam and Mike each have a newspaper round.

The combined weight of their three bags of papers is 21 kg.

Kenny's bag is 600 g lighter than Liam's.

Liam's bag and Mike's bag are exactly the same weight.

What is the weight of each boy's bag?

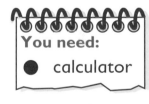

You need:
- calculator

2 Make up a similar problem for a friend to solve.

Co-ordinating patterns

● **Use co-ordinates and translate shapes on grids**

 1 A game at the school fair is guessing where the treasure is buried.

Copy the grid onto RCM 11.

Complete this list of guesses.

A (__ , 5)

B (__ , 4)

C (1, __)

D (__ , 3)

E (__ , __)

F (__ , __)

G (__ , __)

H (__ , __)

You need:
● RCM 11: 10 × 6 co-ordinate grids
● ruler

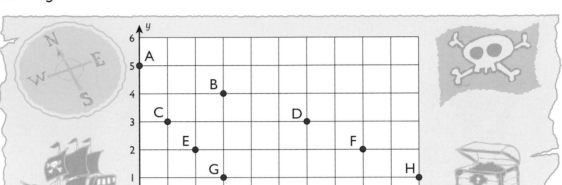

2 Four unclaimed prizes are buried at these points.

Plot the points on the co-ordinate grid.   P1 (3, 3)   P2 (7, 1)   P3 (2, 1)   P4 (8, 5)

 1 On RCM 11, plot these points and join them up in order.

Name the shape you draw.

a   (1, 4) (3, 6) (7, 2) (5, 0)

b   (1, 0) (4, 6) (7, 5) (8, 2)

You need:
● RCM 13, 10 × 6 co-ordinate grid
● ruler

2 In these diagrams you can see one side of a square.

Write your first choice of co-ordinates for points C and D.

Now find a different set of vertices for each shape.

a

A
●
(5,5)

B
●
(5,1)

A ___ B
D ___ C

D ___ C
A ___ B

A ___ D
B ___ C

b

A ● —————— B ●
(3,5)   (8,5)

c

A ● —————— B ●
(1,6)   (5,6)

3 This table shows the co-ordinates for the corresponding vertices of the shape in Grid 1.

	x-axis	y-axis
Shape A	1	5
Shape B	4	4

The shape is translated 3 to the right, then 1 down. Copy and complete the table for the corresponding vertices in shapes C and D.

4 For each of the Grids 2 and 3:

a draw and complete a table for the corresponding vertices.

b write the distance and direction of the translation, e.g. 3 to the right, then 1 down.

c write the next two rows for each table.

5 Explain the connection between the translated shapes and the pattern of corresponding vertices.

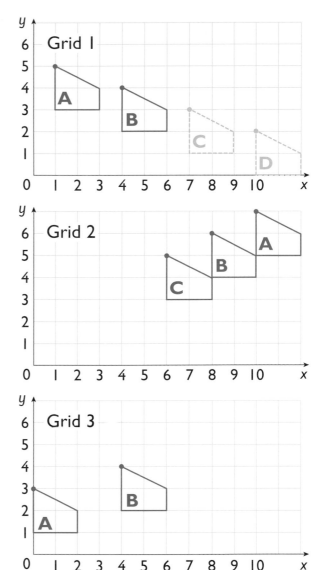

1 Translate the shape using the rule: 2 to the right, then 2 up. Then translate the shape using the rule: 3 to the right, then 1 down.

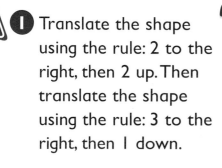

You need:
- 1 cm squared paper
- ruler
- coloured pencil

2 Continue the translations, alternating between each one until you have filled the grid.

3 Use one colour only for each translation.

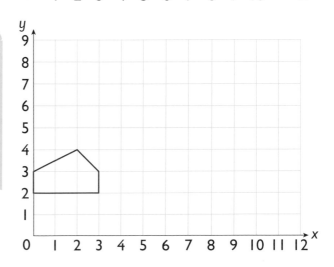

Measuring angles

- Use a protractor to measure angles
- Check the sum of the angles in a triangle

These crisscross lines are fresh ski tracks in the snow.
Measure and record to the nearest degree the size
of each marked angle.

You need:

- protractor

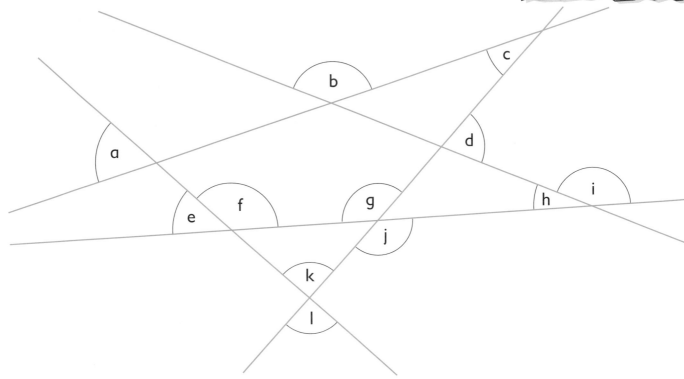

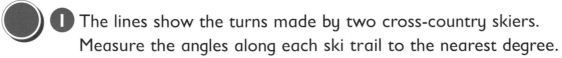

1 The lines show the turns made by two cross-country skiers.
Measure the angles along each ski trail to the nearest degree.

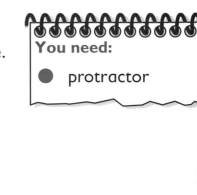

You need:

- protractor

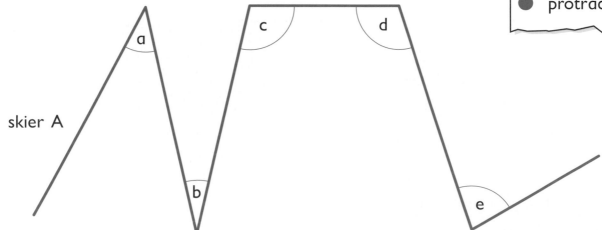

skier A

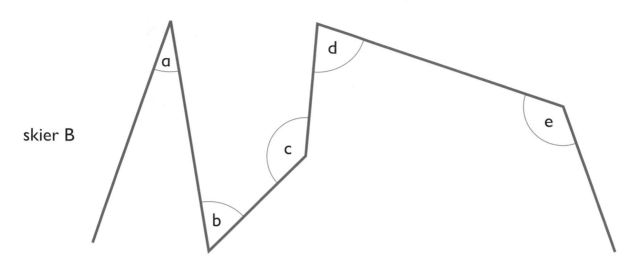

skier B

2 Measure and add the angles for each triangle.

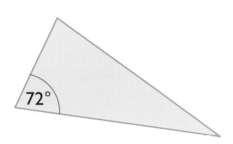

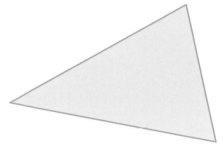

a 72° + ___° + ___° = ___°

b ___° + ___° + ___° = ___°

 ## Chalet roof puzzle

1 a Draw a large triangle for the roof of the chalet.

b Mark the midpoints of the sides and join them up.

c Measure all the angles in each of the 4 small triangles.

d Write down what you notice.

2 What happens if you draw a different triangle? Will it still work? Investigate.

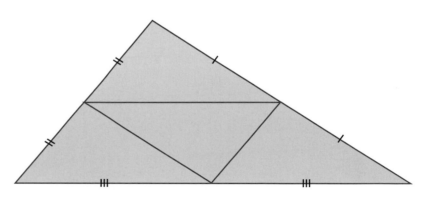

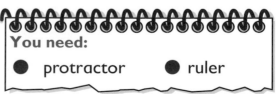

You need:
● protractor   ● ruler

81

Drawing angles

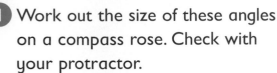

● **Use a protractor to draw angles**

You need:
● protractor

1 Work out the size of these angles on a compass rose. Check with your protractor.

Remember

The amount of turn between N and NE is 45°.

a
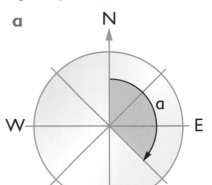

b
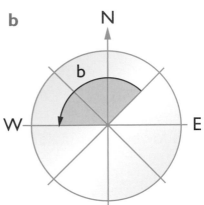

c
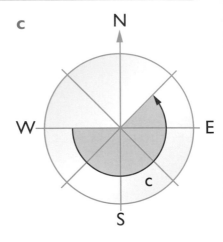

2 Write the direction in which you will face after turning through these reflex angles:

a Face west. Turn clockwise through 270°.

b Face north. Turn clockwise through 225°.

c Face east. Turn clockwise through 315°.

d Face SW. Turn anticlockwise through 270°.

e Face NE. Turn anticlockwise through 315°.

Example

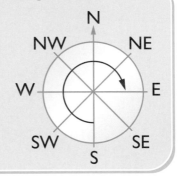

Face south.

Turn clockwise through 270°.

You now face east.

1 Copy this table, then write the angles in the correct column:

acute	right	obtuse	straight	reflex
57°				

a 75°   c 205°   e 134°   g 36°   i 190°

b 157°   d 340°   f 180°   h 90°   j 275°

You need:
● protractor
● ruler

2 Draw and label angles of these sizes:

a 65°   c 108°   e 149°

b 137°   d 42°   f 161°

3 Make accurate drawings
of these angles.

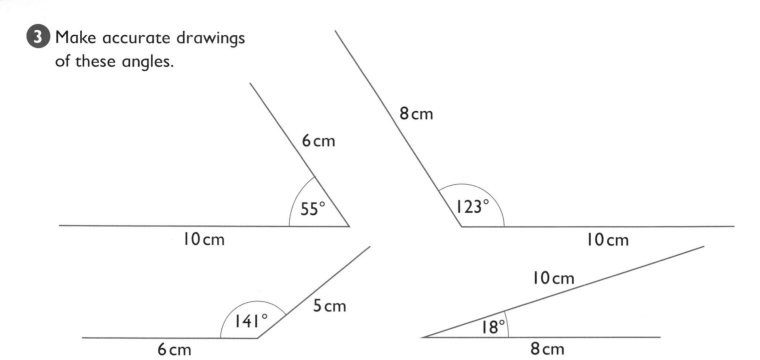

4 Construct these triangles.

a Draw a line AB 10 cm long.

- At A draw an angle of 40°.
- At B draw an angle of 65°.
- Extend the lines at A and B to meet at C.
- Measure the size of the angle at C.

b Draw a line PQ 8 cm long.

- Draw an angle of 57° at P and Q.
- Extend the lines at P and Q to meet at R.
- Measure the size of the angle at R.

1 Use the edge of your protractor to construct 5 semicircles from 0° to 180°.

Join the points at 0° and 180° to complete each semicircle.

You need:
- protractor
- ruler

2 For each semicircle:

a Mark a point on the circumference.

b Join the point to each end of the diameter line.

c Measure the angle at the circumference.

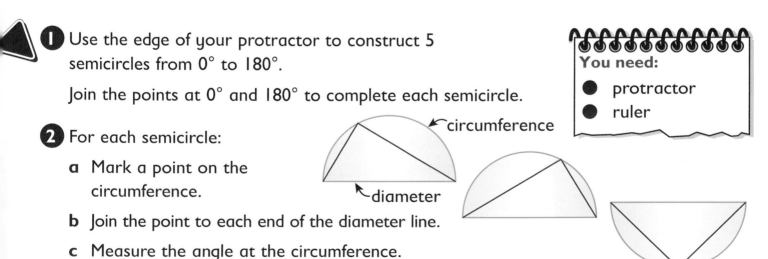

3 Make a general statement about the angle at the circumference of a semicircle.

Calculating angles

● **Know the angle sum of a triangle is 180° and the sum of angles around a point is 360°**

 These lines show the route of a bee from flower to flower.

Use your protractor to measure the size of the marked angles.

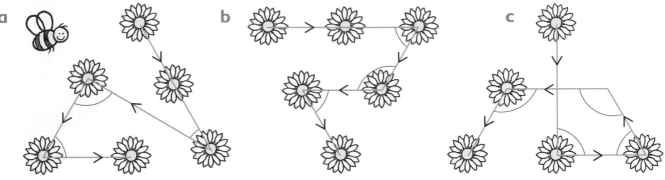

a
b
c

2 This trail and some helpful clues were left by a garden snail.

a Calculate the size of each missing angle.

b Now check your answers with your protractor.

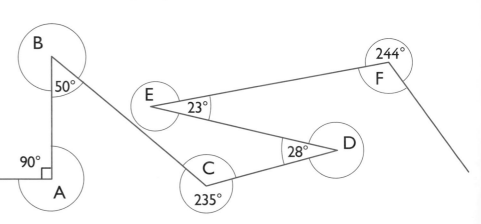

1 a PQR is a straight line. Measure and record the size of angles S and T.

b One angle is 35°. What is the sum of angles S and T?

c Compare with your measurements. How accurate were you?

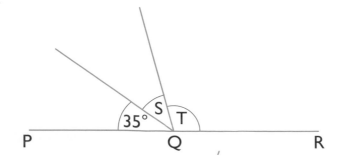

2 a Measure and record the size of angles A, B and C.

b If DEF is a straight line, what should be the total of angles A + B + C?

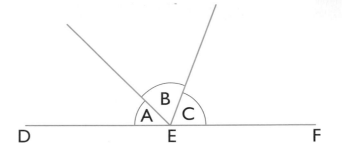

3 First measure, then check by working out, the marked angles on these sails.

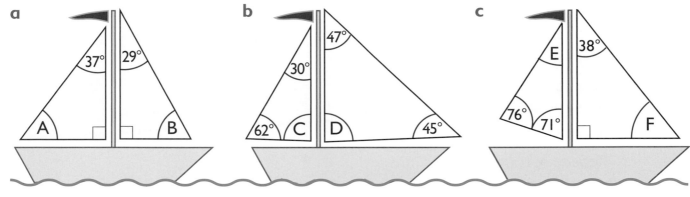

a 37° 29° A B

b 47° 30° 62° C D 45°

c E 38° 76° 71° F

4 Name and calculate the sizes of the shaded angles.

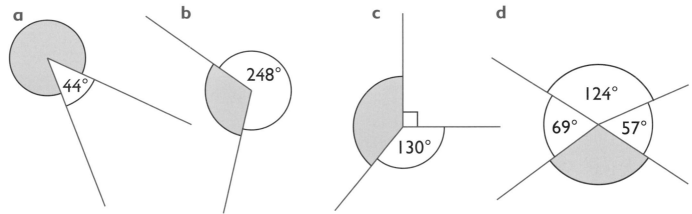

a 44°

b 248°

c 130°

d 124° 69° 57°

1 Construct a regular pentagon which has sides of 6 cm and angles of 108°.

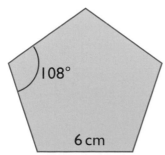

108°

6 cm

2 Construct a regular hexagon which has sides of 5·5 cm and angles of 120°.

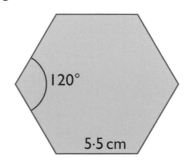

120°

5·5 cm

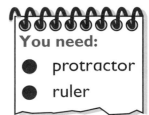
You need:
- protractor
- ruler

Rotate the shape

● Recognise where a shape will be after a rotation
through 90° or 180° about one of its vertices

Make a pattern by rotating a shape through 90°.

You need:
- small square of card
- ruler
- scissors

a On a square of card, make cuts on
2 adjacent sides.

b Draw 2 intersecting perpendicular lines.

c Place the card in the 1st
position and draw round it.

d Rotate the card anticlockwise
through 90°, into the 2nd position,
and draw round it.

e Repeat rotations of 90° into
the 3rd and 4th positions.

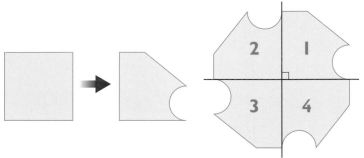

When you rotate a T-shape clockwise through 90°
about the point A, three times, you make this shape.

1 a Make templates of
these shapes on 1 cm
squared paper.
Mark the point A.

b For each template,
record on 1 cm
squared paper,
3 rotations through 90° about the point A.

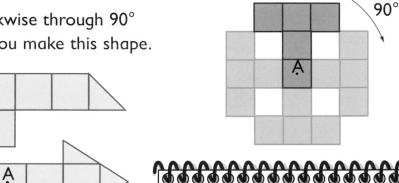

You need:
- 1 cm squared paper
- ruler
- RCM 13, 12 × 12
  co-ordinate grids

2 Copy these shapes on to 1 cm squared paper.

Complete each
drawing by
rotating the
shape through
180° about
the dot.

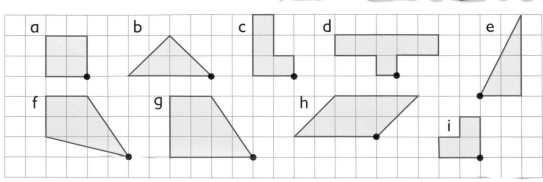

3 **a** Copy these shapes on to 12 × 12 co-ordinate grids or RCM 12.

b For each shape draw three clockwise rotations of 90° about point (6, 6).

c Write the co-ordinates of vertices B and Q for each rotation.

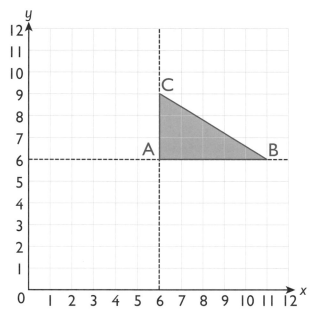

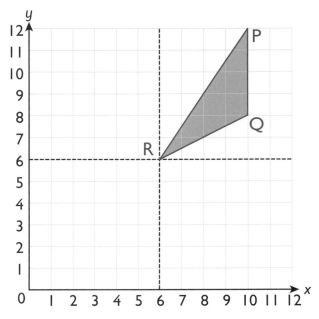

1 Using a 12 × 12 co-ordinate grid or RCM 12, rotate the kite KLMN 90° clockwise about the point O.

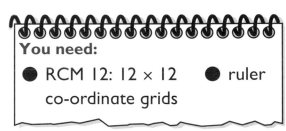

You need:

● RCM 12: 12 × 12 co-ordinate grids

● ruler

2 Enter the co-ordinates of the vertices of each kite in a table.

vertex	starting position	90° clockwise rotation		
		1st	2nd	3rd
K	(7, 7)	(7, 5)		
L	(11, 8)			
M				
N				

3 Write about any patterns you notice.

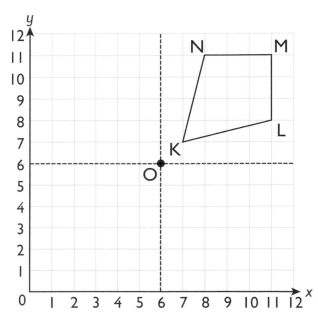

Using the written method

 1 Choose two numbers from below.

Write out either an addition or a subtraction calculation.

Make up **8** calculations altogether.

You should do about half addition and half subtraction calculations.

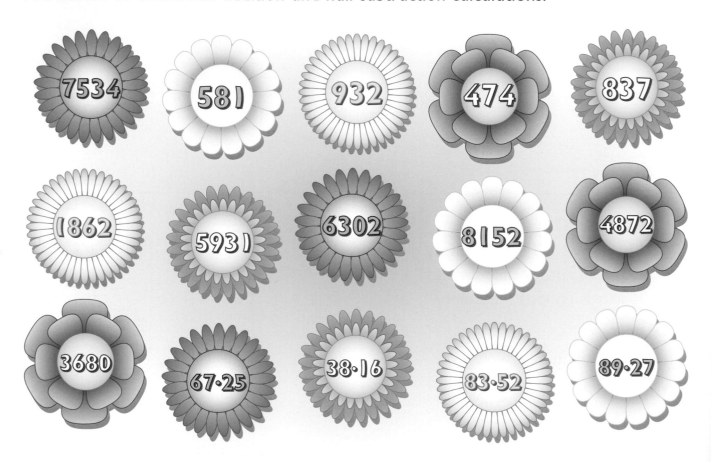

7534   581   932   474   837

1862   5931   6302   8152   4872

3680   67·25   38·16   83·52   89·27

2 Look at these calculations. Copy them out and put in the missing digits.

a
```
    □ 5 4
+   5 □ □
─────────
    7 9 2
```

b
```
    8 □ □
+   □ 0 9
─────────
  1 2 7 4
```

c
```
    9 □ 2
−   □ 2 □
─────────
    5 1 5
```

1 Choose two numbers from below. Write out either an addition or a subtraction calculation.

Make up 10 calculations altogether.

You should do about half addition and half subtraction calculations.

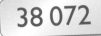

You need:

● a calculator

38 072	34 832	12 825	17 037
44 193	10 472	5839	9034
3658	7282	6219	459·43
272·38	183·35	3842·81	2941·74

Now check your answers with a calculator.

2 Copy out these calculations and work out the missing numbers.

The digits 2 3 5 7 9 0 have been used once in every calculation.

a
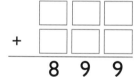
```
   □ □ □
 + □ □ □
─────────
   8 9 9
```

d
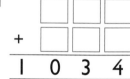
```
   □ □ □
 + □ □ □
─────────
 1 0 3 4
```

g
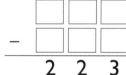
```
   □ □ □
 − □ □ □
─────────
   2 2 3
```

b
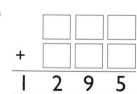
```
   □ □ □
 + □ □ □
─────────
 1 2 9 5
```

e
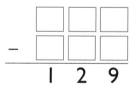
```
   □ □ □
 − □ □ □
─────────
   1 2 9
```

h
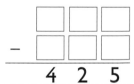
```
   □ □ □
 − □ □ □
─────────
   4 2 5
```

c
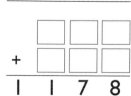
```
   □ □ □
 + □ □ □
─────────
 1 1 7 8
```

f
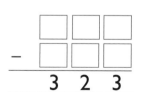
```
   □ □ □
 − □ □ □
─────────
   3 2 3
```

i
```
   □ □ □
 + □ □ □
─────────
 1 5 0 2
```

My answer is 14 483.

Can you find 6 four-digit add four-digit calculations that equal this number?

Write them out as written calculations and show your working.

Inverse problems

● **Solve multi-step problems, and problems involving decimals**

 Work out the following problems. For each problem you must show your method of working out.

Write down the inverse calculation you use to solve the problem.

a I am thinking of a number. I add 85 to it and my answer is 147. What was my number?

b I am thinking of a number. I add 153 to it and my answer is 241. What was my number?

c I am thinking of a number. I add 3·2 to it and my answer is 6·3. What was my number?

d I am thinking of a number. I subtract 84 from it and my answer is 132. What was my number?

e I am thinking of a number. I subtract 163 from it and my answer is 225. What was my number?

f I am thinking of a number. I subtract 7·2 from it and my answer is 4·1. What was my number?

Example

I am thinking of a number.
I add 53 to it and my answer is 92.
What was my number?

Problem ☐ + 53 = 92

Inverse calculation 92 − 53 = ☐

Example

Problem ☐ − 48 = 87

Inverse calculation 87 + 48 = ☐

 Work out the following problems. For each problem you must show your method of working out.

Write down the inverse calculation you use to solve the problem.

a I am thinking of a number. I add 383 to it and my answer is 431. What was my number?

b I am thinking of a number. I add 5·8 to it and my answer is 9·4. What was my number?

c I am thinking of a number. I add 604 to it and my answer is 749. What was my number?

d I am thinking of a number. I add 7·9 to it and my answer is 9·8. What was my number?

e I am thinking of a number. I add 276 to it and my answer is 863. What was my number?

f I start with 341. When I add 285 and then subtract a number, the answer is 267.
What number did I subtract?

g I am thinking of a number. I subtract 3·2 from it and my answer is 5·8. What was my number?

h I am thinking of a number. I subtract 429 from it and my answer is 174. What was my number?

i I am thinking of a number. I subtract 758 from it and my answer is 481. What was my number?

j I am thinking of a number. I subtract 6·5 from it and my answer is 7·3. What was my number?

k I start with 747. When I add 342 and then subtract a number, the answer is 267. What number did I subtract?

l I start with 601. When I subtract 437 and then add a number, the answer is 731. What number did I add?

 a Joe thinks of a number. He adds 267 and then multiplies it by 2.
The answer he gets is 760. What number did he start with?

b Jamie thinks of a number. He says, 'I multiplied it by itself, and then halved it. The answer is 392.' What was his number?

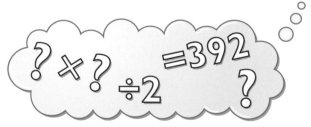

Work it out!

- Choose and use appropriate calculation strategies including a calculator

Work out the answers to these problems.

You need:
- calculator

1 A tea cup holds 240 ml.

   a In a week Mrs Brown drinks 8400 ml of tea. How many cups is that?

   b The tea pot holds 2 litres. How many full cups can be poured out of it?

   c If I drink 4 cups of tea a day, how many cups of tea will I drink in a 7-day week?

2 Chocolate bars come in packs of 8.

£1·84

   a The pack costs £1.84. How much is each bar?

   b I buy 10 packs. How much will that cost me?

   c 4 friends buy 1 pack and they each have 2 chocolate bars. How much does each friend pay?

Work out the answers to these problems.

You need:
- calculator

1 Pitta bread is sold in packs of 6. The bags are packed in boxes.

   a Each box contains 30 packs. How many pitta breads does it hold?

   b If a supermarket has 4 boxes, how many packs do they have? How many pitta breads?

   c If a pack costs 63p, how much does a box cost?

2 At the music shop a DVD costs £6.85 and a CD costs £9.45.

   a Jan bought a DVD and 2 CDs. How much did she spend?

   b Karen bought 2 DVDs and paid with a £20 note. How much change did she get?

   c A school buys 10 DVDs and 10 CDs. They get a 10% reduction. What is their total bill?

3 A coach holds 52 passengers.

a A school of 180 children and staff are
going on an outing.
How many coaches do they need?

b Each coach costs £72 per day.
How much will the school coaches cost?

c The coach company has 12 coaches.
If they are all full, how many people
will there be altogether?

4 A box of matches holds 98 matches.
They come in packs of 10.

a How many matches in a pack?

b If I have 9800 matches, how many boxes do I have?

c If I buy 100 packs, how many matches will I have?

Think of 3 word problems. Work out the answers,
then give them to a friend to solve.

At the shoe
shop I buy …

I bottle holds
450 ml …

Keep it simple!

Simplify these fractions to their simplest form.

Remember, you have to find a number that the numerator and the denominator can both be divided by.

1 a $\dfrac{4}{12}$ f $\dfrac{10}{15}$

b $\dfrac{3}{9}$ g $\dfrac{7}{21}$

c $\dfrac{2}{10}$ h $\dfrac{15}{20}$

d $\dfrac{9}{12}$ i $\dfrac{3}{15}$

e $\dfrac{6}{8}$ j $\dfrac{16}{24}$

2 Using your answers to question **1**, sort all the fractions into equivalent groups.

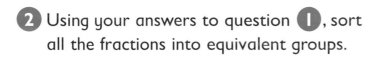

Example
$$\dfrac{4}{12} = \dfrac{3}{9}$$

Choose one number from each box and make a proper fraction.
Simplify that fraction to its simplest form. Do this for 15 fractions.

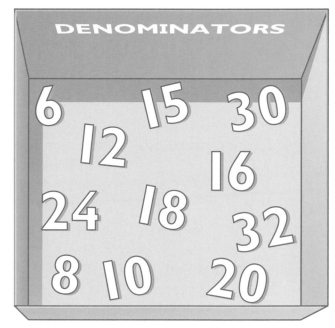

Play this game with a partner.

- Decide whether the larger or the smaller fraction is going to win.

- Both secretly choose a number from each box in the ⬤ section.

- Make a fraction and simplify it.

- Show your fractions to each other.

- The person with the larger or smaller fraction is the winner.

- How will you decide whose fraction is the larger / smaller?

Fractions

● **Order fractions by converting them to fractions with a common denominator**

1 Change these improper fractions to mixed numbers.

a $\frac{15}{8}$ b $\frac{4}{3}$ c $\frac{9}{8}$ d $\frac{13}{6}$ e $\frac{17}{4}$

f $\frac{11}{2}$ g $\frac{16}{5}$ h $\frac{20}{9}$ i $\frac{18}{10}$ j $\frac{121}{100}$

2 Change these mixed numbers to improper fractions.

a $1\frac{3}{6}$ b $1\frac{4}{5}$ c $1\frac{2}{7}$ d $2\frac{4}{8}$ e $2\frac{1}{3}$

f $2\frac{6}{10}$ g $2\frac{8}{9}$ h $2\frac{7}{12}$ i $3\frac{3}{10}$ j $3\frac{18}{100}$

3 Order these fractions, smallest to largest,
by first converting them to equivalent fractions.

a $\frac{1}{2}$ $\frac{3}{8}$ $\frac{1}{4}$ Convert these fractions to eighths.

b $\frac{3}{4}$ $\frac{7}{8}$ $\frac{2}{3}$ Convert these fractions to twenty-fourths.

c $\frac{1}{3}$ $\frac{4}{6}$ $\frac{1}{2}$ Convert these fractions to sixths.

d $\frac{1}{4}$ $\frac{5}{8}$ $\frac{1}{2}$ Convert these fractions to eighths.

e $\frac{2}{5}$ $\frac{3}{4}$ $\frac{6}{10}$ Convert these fractions to twentieths.

f $\frac{7}{12}$ $\frac{1}{4}$ $\frac{2}{3}$ g $\frac{7}{10}$ $\frac{12}{20}$ $\frac{4}{5}$

h $\frac{4}{6}$ $\frac{9}{12}$ $\frac{1}{3}$ i $\frac{7}{9}$ $\frac{2}{3}$ $\frac{14}{18}$

j $\frac{9}{12}$ $\frac{1}{2}$ $\frac{5}{6}$ k $\frac{11}{15}$ $\frac{1}{5}$ $\frac{2}{3}$

Example

$\frac{1}{2}$, $\frac{3}{4}$, $\frac{5}{8}$ → $\frac{4}{8}$, $\frac{6}{8}$, $\frac{5}{8}$

$\frac{1}{2}$, $\frac{5}{8}$, $\frac{3}{4}$

I converted these fractions to eighths. Now I can order them.

1 Order these fractions, smallest to largest.
First convert them to equivalent fractions, then
draw a number line and write the fractions.

a $\frac{5}{8}$, $\frac{3}{4}$, $\frac{10}{16}$

b $1\frac{2}{7}$, $1\frac{1}{3}$, $1\frac{5}{21}$

c $2\frac{5}{6}$, $2\frac{2}{3}$, $2\frac{7}{9}$

d $5\frac{4}{12}$, $5\frac{1}{4}$, $5\frac{1}{3}$

e $6\frac{16}{20}$, $6\frac{3}{5}$, $6\frac{7}{10}$

f $2\frac{6}{7}$, $2\frac{1}{2}$, $2\frac{11}{14}$

g $4\frac{18}{25}$, $4\frac{3}{5}$, $4\frac{9}{10}$

h $7\frac{8}{10}$, $7\frac{73}{100}$, $7\frac{11}{20}$

i $9\frac{2}{8}$, $9\frac{9}{32}$, $9\frac{1}{2}$

j $12\frac{23}{30}$, $12\frac{4}{6}$, $12\frac{1}{5}$

Example

$\frac{1}{2}$, $\frac{4}{6}$, $\frac{1}{3}$

$\frac{1}{3} = \frac{2}{6}$

$\frac{1}{2} = \frac{3}{6}$

0, $\frac{1}{3}$, $\frac{1}{2}$, $\frac{4}{6}$, 1

2 What number is halfway between these numbers?

a $3\frac{1}{4}$ and $3\frac{1}{2}$

b $5\frac{1}{3}$ and $5\frac{2}{3}$

c $3\frac{2}{5}$ and $3\frac{3}{5}$

d $1\frac{4}{5}$ and $1\frac{4}{6}$

e $6\frac{8}{10}$ and $6\frac{2}{5}$

f $9\frac{6}{12}$ and $9\frac{2}{3}$

g $4\frac{2}{8}$ and $4\frac{6}{16}$

h $8\frac{6}{10}$ and $8\frac{4}{20}$

i $1\frac{4}{9}$ and $1\frac{2}{3}$

j $2\frac{3}{5}$ and $2\frac{1}{3}$

Work in pairs.

- Draw one number line on a piece of paper.

- Each person writes a fraction.

- Convert both fractions to a fraction with a common denominator.

- Put them on the number line.

- Repeat four more times.

Remember

Think carefully where the fractions go on the number line!

Find the fraction

● **Find fractions of whole number quantities**

Find the fractions of these numbers.

You may want to use a calculator for some calculations.

Record your working out.

a $\frac{1}{4}$ of 440

b $\frac{1}{2}$ of 360

c $\frac{1}{3}$ of 312

d $\frac{1}{5}$ of 120

e $\frac{1}{4}$ of 96 , $\frac{3}{4}$ of 96

f $\frac{1}{5}$ of 250 , $\frac{3}{5}$ of 250

g $\frac{1}{6}$ of 60 , $\frac{5}{6}$ of 60

h $\frac{1}{10}$ of 160 , $\frac{4}{10}$ of 160

i $\frac{1}{100}$ of 400 , $\frac{7}{100}$ of 400

j $\frac{1}{8}$ of 48 , $\frac{6}{8}$ of 48

k $\frac{1}{4}$ of 8000 , $\frac{3}{4}$ of 8000

l $\frac{1}{3}$ of 618 , $\frac{2}{3}$ of 618

Example

$\frac{3}{5}$ of 200

$\frac{1}{5}$ of 200 = 40

(200 ÷ 5 = 40)

$\frac{3}{5}$ of 200 = 120

(40 × 3 = 120)

Find the fractions of these numbers and quantities.

You may want to use a calculator for some calculations.

Record all your working out.

You need:
● calculator
  (optional)

a $\frac{3}{10}$ of 80, £50, 120

f $\frac{2}{5}$ of 400, £85, 550

b $\frac{46}{100}$ of 3000, 2600, £1500

g $\frac{8}{10}$ of 410, 12 m, 1100

c $\frac{5}{6}$ of 600, 120 km, 9 m

h $\frac{6}{100}$ of 60 m, 4000, £5

d $\frac{3}{4}$ of 51, 900, 266

i $\frac{2}{6}$ of 31, 3840, £96

e $\frac{2}{8}$ of 448, 112, 3200 m

j $\frac{5}{8}$ of £850, 1608, 592

Work out the fractions of the following measurements:

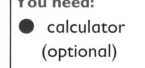

You need:
● calculator
  (optional)

a $\frac{3}{10}$ of 2 m in centimetres?

d $\frac{7}{10}$ of 5 m in centimetres?

b $\frac{23}{100}$ of 4 kg in grams?

e $\frac{55}{100}$ of 7 litres in millilitres?

c $\frac{7}{1000}$ of 1 m in millimetres?

f $\frac{3}{4}$ of 8 km in metres?

Percentage wheels

Find the percentages of these amounts. Use the answer to the first percentage to help you work out the second.

Record your calculations.

Example

50% of £300

£300 ÷ 2 = £150

25% of £300

£150 ÷ 2 = £75

a 50% of £800,   25% of £800

b 50% of £24,   25% of £24

c 50% of £620,   25% of £620

d 50% of £46,   25% of £46

e 10% of £70,   20% of £70

f 10% of £140,   20% of £140

g 10% of £300,   20% of £300

h 10% of £20,   40% of £20

i 10% of £480,   40% of £480

j 10% of £12,   40% of £12

Work out the percentages of the number in the centre of the wheel.

Record your method.

a

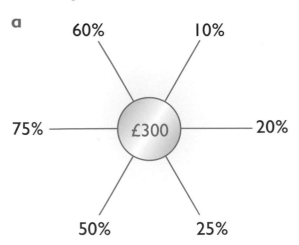

60%   10%

75% —— £300 —— 20%

50%   25%

b

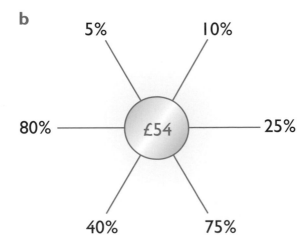

5%   10%

80% —— £54 —— 25%

40%   75%

c
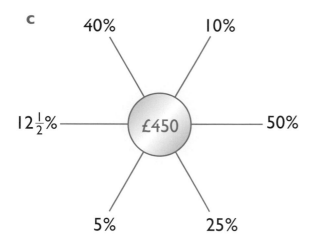

40% 10%

$12\frac{1}{2}$% — £450 — 50%

5% 25%

d
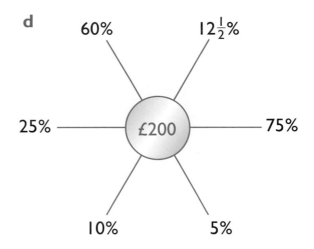

60% $12\frac{1}{2}$%

25% — £200 — 75%

10% 5%

e
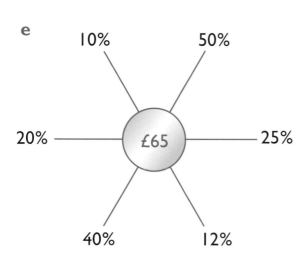

10% 50%

20% — £65 — 25%

40% 12%

f
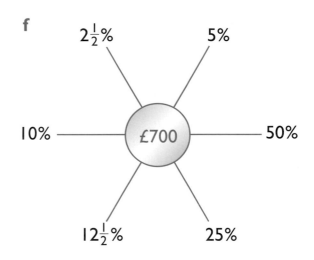

$2\frac{1}{2}$% 5%

10% — £700 — 50%

$12\frac{1}{2}$% 25%

VAT is a tax that is added on to most things we buy.

At the moment VAT is always $17\frac{1}{2}$% of the price.

Work out a way to find $17\frac{1}{2}$% of these prices:

Show all your working.

a £40

b £75

c £15

d £86

e £32

Shop percentages

● **Find percentages of whole number quantities**

The stationery shop is going to close down in 5 days.

Over the next week everything must be sold.

On Monday the prices will be reduced by 10%.

On Tuesday the prices will be reduced by 20%.

On Wednesday the prices will be reduced by 30%.

On Thursday the prices will be reduced by 40%.

On Friday the prices will be reduced by 50%.

Work out the prices of the following items
over the next week.

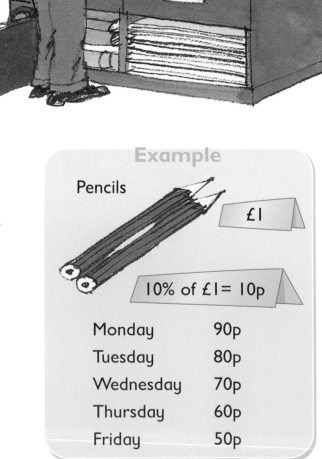

£5

£2

£1.80

£3

£2.60

£4

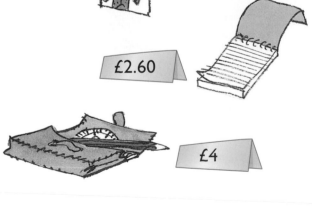

Example

Pencils

£1

10% of £1 = 10p

Monday	90p
Tuesday	80p
Wednesday	70p
Thursday	60p
Friday	50p

1 The clothing shop is going to close down in 5 days. Over the next week everything must be sold.

On Monday the prices will be reduced by 10%.

On Tuesday the prices will be reduced by 15%.

On Wednesday the prices will be reduced by 30%.

On Thursday the prices will be reduced by 50%.

On Friday the prices will be reduced by 75%.

Work out the prices of the following items for each day over the next week.

a £15 – T-shirt

b £28 – Shirt

c £45 – Jacket

d £32 – Dress

e £50 – Trousers

2 At the supermarket you get extra free. Work out what percentage is free.

a 440 g for the price of 400 g

b 300 g for the price of 250 g

c 100 g for the price of 75 g

d 9 l for the price of 8 l

e 21 biscuits for the price of 20

 You own a shop. You are going to have a sale. Work out the original prices for 10 items and the percentage you are going to reduce them by.

Ask a friend to calculate your sale prices.

Fraction and decimal clouds

 Use all the digits in the stars to make
a fraction and its decimal equivalent.

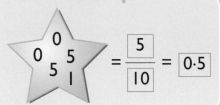

$$\frac{5}{10} = \boxed{0.5}$$

a

b

c

d

e

f

g

h

i

 Use all the digits in the clouds to make
a fraction and its decimal equivalent.

Example

$$\frac{985}{1000} = \boxed{0} \cdot \boxed{985}$$

a

b

c

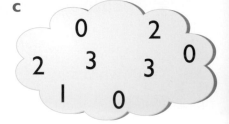

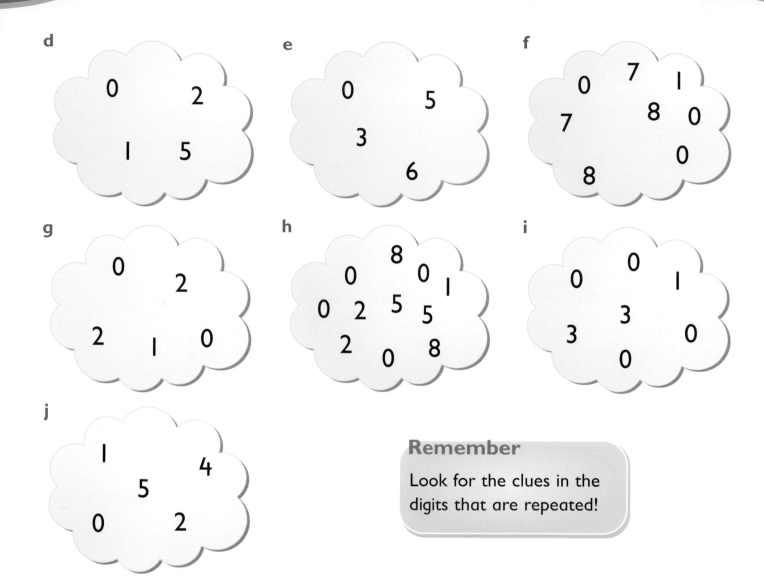

d

0   2
1   5

e

0   5
3
6

f

7   1
0
7   8   0
8   0

g

0   2
2   1   0

h

8   0
0   1
0   2   5   5
2
0   8

i

0
0   1
3
3   0
0

j

1
4
5
0   2

Remember

Look for the clues in the
digits that are repeated!

 Choose the decimal that is equal to the fraction.

a $\frac{193}{100}$ 1·93   0·193   10·193   19·13

b $\frac{47}{1000}$ 0·47   47·00   0·047   4·07

c $\frac{2}{1000}$ 0·02   0·2   0·002   0·0002

d $\frac{673}{1000}$ 0·763   0·673   6·73   0·0673

e $\frac{405}{1000}$ 0·045   0·504   4·405   0·405   0·004

What's the equivalent?

● **Find equivalent percentages, decimals and fractions**

1 Find the decimal equivalent to the following fractions.

a $\frac{1}{2}$ b $\frac{1}{4}$ c $\frac{1}{5}$

d $\frac{1}{8}$ e $\frac{1}{10}$ f $\frac{1}{100}$

> Remember you divide 1 by the denominator to find the decimal equivalent.

2 Now find the percentage equivalent.

> Remember you divide 100 by the denominator to find the percentage equivalent.

1 Copy and complete the following tables.

Use a calculator if you need to.

You need:
● calculator (optional)

Example

$\frac{3}{5}$ 0·6

$1 \div 5 = 0·2$

$0·2 \times 3 = 0·6$

Record the calculation you use. Write decimals to two decimal places.

a

fraction	equivalent decimal	calculation
$\frac{1}{5}$		
$\frac{1}{10}$		
$\frac{1}{100}$		
$\frac{1}{8}$		
$\frac{1}{3}$		
$\frac{1}{7}$		
$\frac{1}{6}$		
$\frac{1}{9}$		
$\frac{3}{4}$		

b

fraction	equivalent %	calculation
$\frac{1}{5}$		
$\frac{1}{10}$		
$\frac{1}{100}$		
$\frac{1}{8}$		
$\frac{1}{3}$		
$\frac{1}{7}$		
$\frac{1}{6}$		
$\frac{1}{9}$		
$\frac{3}{4}$		

2 Explain the link between fractions, decimals and percentages.

Find the decimal and percentage equivalents to the following fractions. Estimate the answers before you work them out, then use a calculator.

You need:

● calculator

a $\frac{1}{19}$ **b** $\frac{1}{11}$ **c** $\frac{1}{12}$ **d** $\frac{1}{20}$ **e** $\frac{1}{15}$

What's the equivalent again?

● **Find equivalent percentages, decimals and fractions**

Copy and complete the following table.
Use a calculator if you need to.

1

fraction	calculation to find equivalent decimal	fraction	calculation to find equivalent decimal
$\frac{1}{4}$	$1 \div 4 = 0.25$	$\frac{3}{4}$	$0.25 \times 3 = 0.75$
$\frac{1}{5}$		$\frac{4}{5}$	
$\frac{1}{10}$		$\frac{7}{10}$	
$\frac{1}{8}$		$\frac{5}{8}$	
$\frac{1}{3}$		$\frac{2}{3}$	

You need:
● a calculator (optional)

2 Now find the percentage equivalents.

fraction	calculation to find equivalent percenatge	fraction	calculation to find equivalent percentage
$\frac{1}{4}$	$100 \div 4 = 25\%$	$\frac{3}{4}$	$25\% \times 3 = 75\%$
$\frac{1}{5}$		$\frac{4}{5}$	
$\frac{1}{10}$		$\frac{7}{10}$	
$\frac{1}{8}$		$\frac{5}{8}$	
$\frac{1}{3}$		$\frac{2}{3}$	

Copy and complete the following tables.

Use a calculator if you need to.

Record the calculation you use. Write decimals to two decimal places.

a

fraction	equivalent decimal	calculations
$\frac{3}{4}$		
$\frac{2}{5}$		
$\frac{2}{3}$		
$\frac{3}{8}$		
$\frac{5}{8}$		
$\frac{4}{10}$		
$\frac{4}{5}$		
$\frac{7}{8}$		

b

fraction	equivalent %	calculations
$\frac{3}{4}$		
$\frac{2}{5}$		
$\frac{2}{3}$		
$\frac{3}{8}$		
$\frac{5}{8}$		
$\frac{4}{10}$		
$\frac{4}{5}$		
$\frac{7}{8}$		

You need:
● a calculator
(optional)

Example

$\frac{3}{5}$ 0·6

$1 \div 5 = 0·2$

$0·2 \times 3 = 0·6$

Choose one fraction, one decimal and one percentage for your friend to put in order from smallest to largest. Make it tricky, but remember you must know the answer!

Fractions and division

 1

Roll two 1-6 dice.

Look at the two digits and make a two-digit number.

Write it down.

Then roll one of the dice again.

Divide your two-digit number by the number on the die.

Write down the answer.

Write the remainder as a fraction.

Make up ten questions in this way.

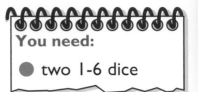

You need:
● two 1-6 dice

Example

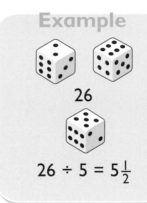

26

$26 \div 5 = 5\frac{1}{2}$

2 Look at the 10 numbers you have made in question **1** .

Add up all the numbers that have the same fraction.

Start by adding any numbers that have half in them, e.g. $3\frac{1}{2} + 5\frac{1}{2} = 9$.

3 Divide these numbers by 2.

6	10	14	24	80

Example

$8 \div 2 = \frac{1}{2}$ of $8 = 4$.

Write the calculation as a division number sentence
and as a fraction number sentence.
Be sure to write down the answer!

4 Divide these numbers by 4.

12	20	32	40	100

Write the calculation as a division number sentence and as a
fraction number sentence. Be sure to write down the answer!

1 Choose one number from each box. Write a division calculation.

Work out the answer as a fraction. Make up ten calculations.

2 Choose 5 numbers. Divide them by 2.

Write the answers in this way, e.g. $12 \div 2 = \frac{1}{2}$ of $12 = 6$.

3 Choose 5 more numbers and divide them by 3. Write the answers in the same way.

 What is the total of all your answers from question **1** in the ⬤ section?

You will have to convert all your answers to a common fraction.

Multiplying fractions

1 Write the multiplication calculation that goes with each of these pictures.

a

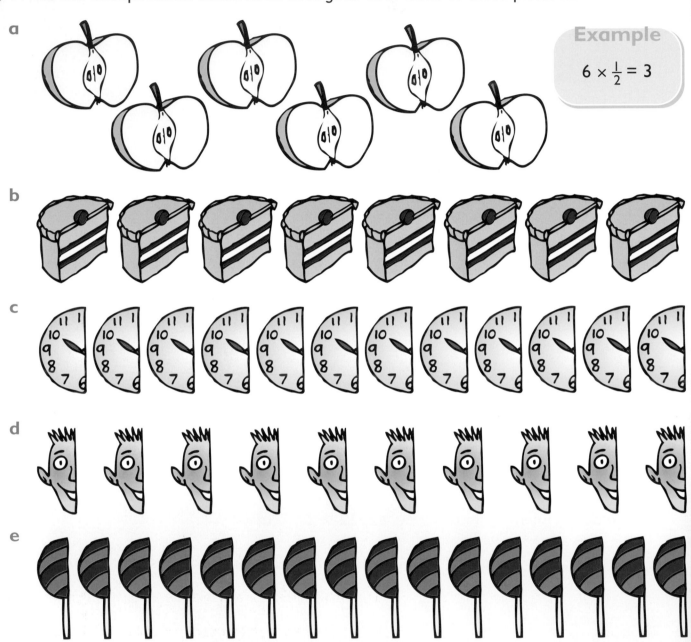

> **Example**
>
> $6 \times \frac{1}{2} = 3$

b

c

d

e

2 Write the other multiplication calculation that goes with each of the questions above.

> **Example**
>
> $\frac{1}{2} \times 6 = 3$

3 Work out these calculations.

a $14 \times \frac{1}{2}$ **b** $20 \times \frac{1}{2}$ **c** $50 \times \frac{1}{2}$

1 Multiply the following numbers by $\frac{1}{2}$.

a 14   **b** 20   **c** 28   **d** 56   **e** 80

2 Multiply the following numbers by $\frac{1}{4}$.

a 20   **b** 32   **c** 40   **d** 100   **e** 400

3 Multiply the following numbers by $\frac{1}{5}$.

a 20   **b** 35   **c** 60   **d** 90   **e** 150

4 Multiply the following numbers by $\frac{1}{8}$.

a 8   **b** 32   **c** 64   **d** 80   **e** 104

5 Choose one of your calculations from each question above and write down how you would check your answer on the calculator.

For each of these calculations write the corresponding division calculation.

a $12 \times \frac{1}{2}$

b $21 \times \frac{1}{3}$

c $30 \times \frac{1}{6}$

d $100 \times \frac{1}{4}$

e $500 \times \frac{1}{5}$

Now make up a word problem for one of your division calculations.

Ratio and proportion patterns

Draw one section of each pattern in your book.

Then work out the ratio of red squares to blue squares.

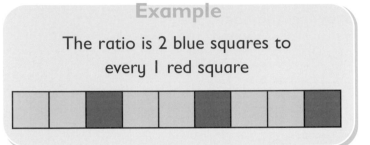

Example

The ratio is 2 blue squares to every 1 red square

You need:
● red and blue pencil

a

The ratio is ⬚ red squares to every ⬚ blue squares.

b

The ratio is ⬚ red squares to every ⬚ blue squares.

c

The ratio is ⬚ red squares to every ⬚ blue squares.

d

The ratio is ⬚ red squares to every ⬚ blue squares.

e

The ratio is ⬚ red squares to every ⬚ blue squares.

What is the ratio and proportion of these patterns?

(First look at how many units there are in each section of the pattern.)

Remember

Use the words:
'to every' for ratio;
'in every' for proportion.

Example

Ratio: 2 white units to every 1 red
Proportion: 2 in every 3 are white

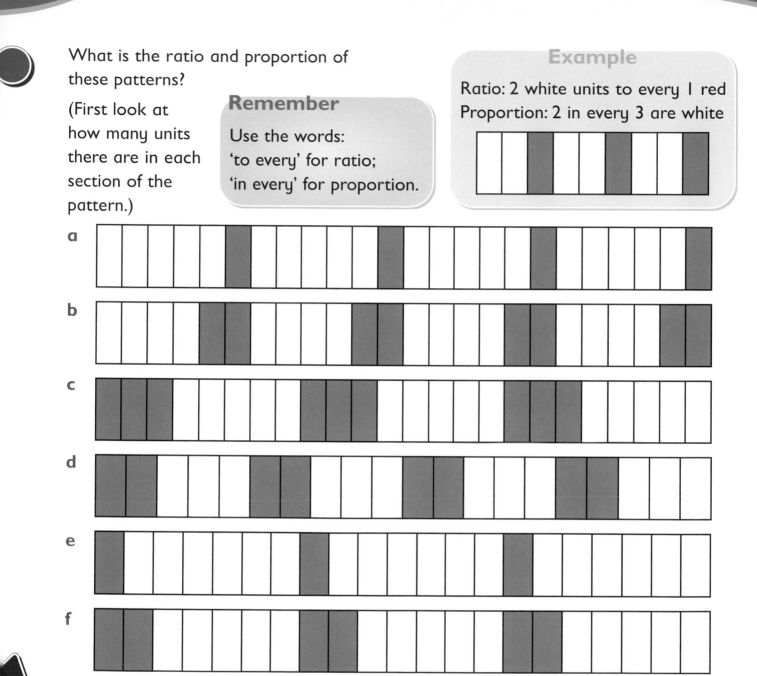

1 Look at the patterns in the ⬤ section and work out how many units would be white and how many red if:

a pattern **a** went on for 30 units

b pattern **b** went on for 30 units

c pattern **c** went on for 40 units

d pattern **d** went on for 35 units

e pattern **e** went on for 42 units

f pattern **f** went on for 49 units

2 Explain how you worked out your answers.

Classroom problems

● Solve simple problems involving ratio and direct proportion

1 Lisa has 12 grapes. She gives 1 to Gavin for every 2 she eats herself.

    a What is the ratio of Lisa's grapes to Gavin's?

    b How many grapes will Lisa eat?

    c How many grapes will Gavin eat?

2 John is looking at his football stickers. For every 1 he wants to keep he has 3 he wants to swap.

    a What is the ratio of stickers he wants to keep to those he wants to swap?

    b If he has 16 stickers, how many stickers does he want to keep?

    c If he has 24 stickers, how many does he want to swap?

3 Ian has counted the flowers in his pots. For every 2 white flowers he has 3 yellow ones. Altogether he has 20 flowers.

    a What is the ratio of white to yellow flowers?

    b How many white flowers are there?

    c How many yellow flowers are there?

1 There are 30 children in Y6. For every 2 girls there is 1 boy.

    a What is the ratio of boys to girls?

    b What proportion of the class is boys?

    c How many girls and how many boys are there?

cake recipe says you need 1 egg to make 5 cakes.
nt to make 25 cakes.

a What is the ratio of eggs to cakes?

b How many eggs will I need?

3 The museum will only allow children to visit if there is a ratio of 1 adult to every 10 children. The school wants to take 50 children.

a How many adults will need to go on the trip?

b What is the proportion of adults on the trip?

4 To make orange paint I mixed 2 pots of red to every 4 pots of yellow. Altogether I used 18 pots of paint.

a What is the ratio of red to yellow paint?

b What proportion of the paint is red?

c How many red pots did I use?

d How many yellow pots did I use?

5 When a Y6 class of 32 children were asked if they preferred English or maths homework, the ratio was 3 : 5. 3 children preferred English to every 5 who preferred maths.

a How many children preferred English homework?

b How many preferred maths homework?

c What proportion of the class preferred English?

d What proportion of the class preferred maths?

Write a word problem with:

a a ratio of 2 : 5

b a ratio of 1 : 6

c a proportion of 1 in every 3

d a proportion of 2 in every 5

Pond borders

a A square garden pond has square slabs with sides 1 m long around the perimeter. Copy these diagrams of square garden ponds on to 1 cm squared paper.

You need:

● 1 cm squared paper
● ruler

b Draw the next two ponds with sides of 3 m and 4 m.

c Copy and complete the table for the number of slabs needed for a pond with up to 4 m sides.

Length of side of pond (m)	1	2	3	4	5
Number of slabs					

d Look for a pattern and use it to find the number of slabs for a square pond with sides of 5 m. Record your results in the table.

1 The DIY centre sells square ponds and square slabs to surround them.

Customer 1 has built a square pond with sides of 8 m.

How many square slabs, with sides of 1 m should he buy?

a On squared paper draw diagrams of ponds with sides of 1 m, 2 m, and 3 m.

Surround each pond with 1 m square slabs.

You need:

● 1 cm squared paper
● ruler
● calculator

Draw diagrams when it helps.

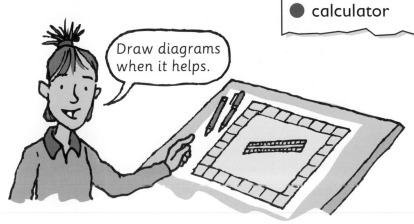

b Copy and complete the table.

Length of side of pond (m)	1	2	3	4	5
Number of slabs					

c Look for a pattern in the table and write it as a rule.

d Use the rule to find the number of slabs the customer needs to buy for his 8 m pond.

e 1 m square slabs cost £7.95 each. Find the cost of the customer's order.

2 Customer 2 has built a rectangular pond with sides of 4 m by 3 m. She is planning to lay square slabs with sides 1 m long around the perimeter of her pond.

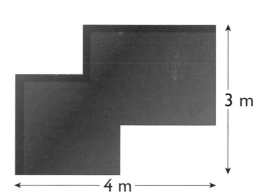

3 m

4 m

a How many 1 m square slabs will she need?

b At £7.95 per slab, find the cost of her order.

3 Customer 3 has designed a pond which has an irregular shape.

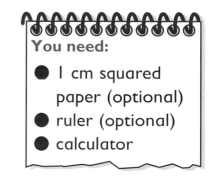

3 m

4 m

a How many square 1 m slabs will he need?

b He chooses a more expensive square slab at £8.49 each. How much will he pay for his slabs?

1 The DIY store sells square slabs with sides of 50 cm.

How many slabs would Customer 3 need to buy to cover the same area as the 1 m slabs surrounding his irregular-shaped pond?

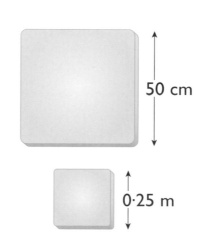

50 cm

0·25 m

You need:
- 1 cm squared paper (optional)
- ruler (optional)
- calculator

2 What if he chooses slabs with sides of 0·25 m?

More slab patterns

● **Draw a diagram and record in a table the steps needed to solve the problem**

1 On 1 cm squared paper draw diagrams to represent these slabs.

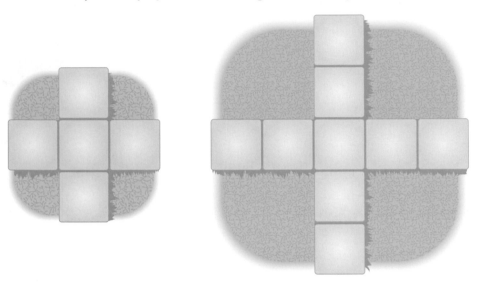

You need:
● 1 cm squared paper

2 Draw the next 3 diagrams to continue the pattern.

3 Copy and complete the table to show your results.

Number of slabs on each arm	1	2	3	4	5		
Total number of slabs							

4 Without drawing a diagram, find the next 2 entries in the table.

5 Explain in words the rule connecting the total number of slabs with the number of slabs on each arm.

A school held a competition for the design of a new path leading from the school to the front gate. This is the winning child's entry.

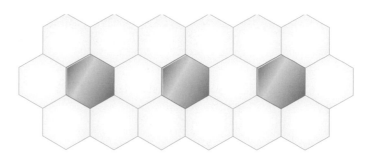

The winning design shows the first 3 terracotta slabs in the pattern.

Each terracotta hexagonal paving slab is surrounded by cream-coloured slabs.

The design uses 100 terracotta paving slabs.

You need:
- 1 cm hexagonal grid paper
- coloured pencils
- calculator (optional)

1 How many cream-coloured hexagonal slabs will be needed to complete the path?

Follow these steps:

- Draw the diagrams for the first 4 terracotta slabs.
- Make a table.
- Look for a pattern.
- Write a formula for the number of cream (C) and terracotta (T) slabs.
- Test your formula.
- Use the formula to answer the question.

2 Find the cost of paving slabs for the new path.

Cost per slab: Cream - £5.99
Terracotta - £6.25

This design won 2nd prize in the new path competition.

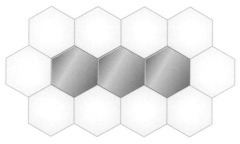

You need:
- 1 cm hexagonal grid paper

Find a formula that the school could use to decide the number of cream and terracotta slabs needed for any length of path of this design.

Intersecting lines

● **Draw a diagram and record in a table the steps needed to solve a puzzle**

Use circles 5 and 6 on your RCM.

Complete the two mystic rose patterns.

Draw straight lines to join each dot to every other dot.

a 10-dot mystic rose

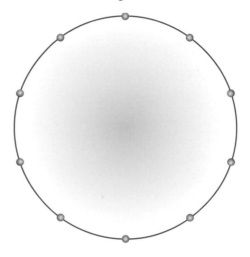

b 12-dot mystic rose

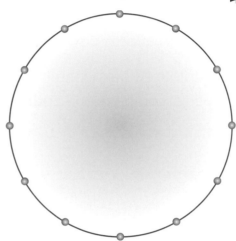

New Homes
in Silverknowes Estate

Every house is:
• on one side of a straight road
• at an equal distance from its neighbour

Straight lines of underground pipes connect each house to:
• the Gas works
• the Water Tower

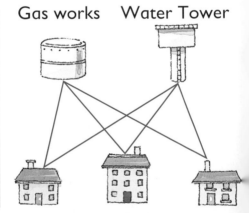

Gas works   Water Tower

In this diagram there are 3 houses and 3 intersections where the gas and water pipes cross over.

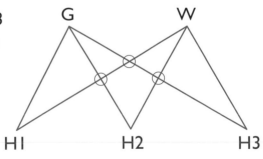

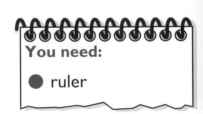

1 **a** How many intersections are there for a road of 4 houses? Copy this diagram.

b What if there are 5 houses? 6 houses? 7 houses?

Draw diagrams to help you.

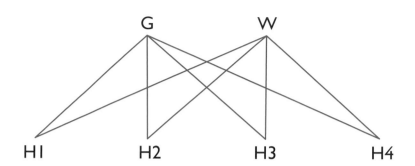

2 Copy and complete this table.

Number of houses	Number of intersections
2	
3	3
4	
5	
6	
7	

3 Predict the number of intersections for a road of 10 houses.

Connecting flights

This diagram shows connecting UK flights for 5 airports.

You can fly from one airport to any other airport.

You have a choice of 10 connecting routes.

a How many connecting flights are there for 8 airports?

b 10 airports?

c Any number of airports worldwide?

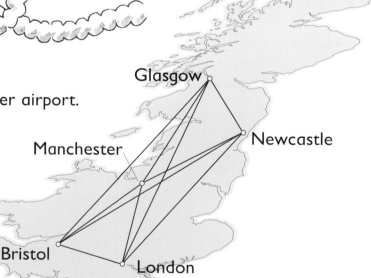

Maths Facts

Problem solving

The seven steps to problem solving

1 Read the problem carefully. **2** What do you have to find?

3 What facts are given? **4** Which of the facts do you need?

5 Make a plan. **6** Carry out your plan to obtain your answer. **7** Check your answer.

Number

Positive and negative numbers

−10 −9 −8 −7 −6 −5 −4 −3 −2 −1 0 1 2 3 4 5 6 7 8 9 10

Place value

1000	2000	3000	4000	5000	6000	7000	8000	9000
100	200	300	400	500	600	700	800	900
10	20	30	40	50	60	70	80	90
1	2	3	4	5	6	7	8	9
0·1	0·2	0·3	0·4	0·5	0·6	0·7	0·8	0·9
0·01	0·02	0·03	0·04	0·05	0·06	0·07	0·08	0·09
0·001	0·002	0·003	0·004	0·005	0·006	0·007	0·008	0·009

Fractions, decimals and percentages

$\frac{1}{100} = 0.01 = 1\%$ $\frac{2}{100} = \frac{1}{50} = 0.02 = 2\%$ $\frac{5}{100} = \frac{1}{20} = 0.05 = 5\%$

$\frac{10}{100} = \frac{1}{10} = 0.1 = 10\%$ $\frac{1}{8} = 0.125 = 12.5\%$ $\frac{20}{100} = \frac{1}{5} = 0.2 = 20\%$

$\frac{25}{100} = \frac{1}{4} = 0.25 = 25\%$ $\frac{1}{3} = 0.333 = 33\frac{1}{3}\%$ $\frac{50}{100} = \frac{1}{2} = 0.5 = 50\%$

$\frac{2}{3} = 0.667 = 66\frac{2}{3}\%$ $\frac{75}{100} = \frac{3}{4} = 0.75 = 75\%$ $\frac{100}{100} = 1 = 100\%$

Number facts

Multiplication and division facts

	×1	×2	×3	×4	×5	×6	×7	×8	×9	×10
×1	1	2	3	4	5	6	7	8	9	10
×2	2	4	6	8	10	12	14	16	18	20
×3	3	6	9	12	15	18	21	24	27	30
×4	4	8	12	16	20	24	28	32	36	40
×5	5	10	15	20	25	30	35	40	45	50
×6	6	12	18	24	30	36	42	48	54	60
×7	7	14	21	28	35	42	49	56	63	70
×8	8	16	24	32	40	48	56	64	72	80
×9	9	18	27	36	45	54	63	72	81	90
×10	10	20	30	40	50	60	70	80	90	100

Tests of divisibility

2 The last digit is 0, 2, 4, 6 or 8.

3 The sum of the digits is divisible by 3.

4 The last two digits are divisible by 4.

5 The last digit is 5 or 0.

6 It is divisible by both 2 and 3.

7 Check a known near multiple of 7.

8 Half of it is divisible by 4 *or*
The last 3 digits are divisible by 8.

9 The sum of the digits is divisible by 9.

10 The last digit is 0.

Calculations

Addition

Whole numbers
Example: 6845 + 5758

```
  6845              6845
+ 5758            + 5758
 11 000            12 603
  1 500              ₁ ₁ ₁
     90
     13
 12 603
     ₁
```

Decimals
Example: 26.48 + 5.375

```
  26.48            26.48
+  5.375          + 5.375
 20.000           31.855
 11.000             ₁ ₁
  0.700
  0.150
  0.005
 31.855
```

Subtraction

Whole numbers
Example: 7845 − 2367

```
  7845      or              700      130      15
− 2367             700      140       5
    33 → 2400      7000 + 800 + 40 + 5
  5445 → 7845    − 2000 + 300 + 60 + 7
  5478             5000 + 400 + 70 + 8
```

```
      7 ¹3¹5
      7̶8̶4̶5̶
    − 2367
      5478
```

Decimals
Example: 639.35 − 214.46

```
  639.35    or
− 214.46
   00.54 → 215
  424.35 → 639.35
  424.89
```

```
      8 ¹2¹5
     6̶3̶9̶.̶3̶5̶
    − 214.46
      424.89
```

Multiplication

Whole numbers
Example: 5697 × 8

×	8
5000	40000
600	4800
90	720
7	56
	45576

```
  5697
 ×   8
 40000  (8 × 5000)
  4800  (8 × 600)
   720  (8 × 90)
    56  (8 × 7)
 45576
     ₁
```

```
  5697
 ×   8
 45576
   ₅ ₇ ₅
```

Decimals
Example: 865.56 × 7

×	7
800	5600
60	420
5	35
0.50	3.5
0.06	0.42
	6058.92

```
  865.56
 ×     7
  5600    (7 × 800)
   420    (7 × 60)
    35    (7 × 5)
   3.5    (7 × 0.50)
  0.42    (7 × 0.06)
 6058.92
     ₁
```

```
  865.56
 ×     7
 6058.92
   ₄ ₃ ₃ ₄
```

Whole numbers
Example: 364 × 87

×	80	7
300	24000	2100
60	4800	420
4	320	28

```
26100
 5220
  348
31668
```

```
   364
 ×  87
 24000  (300 × 80)
  4800  (60 × 80)
   320  (4 × 80)
  2100  (300 × 7)
   420  (60 × 7)
    28  (4 × 7)
 31668
   ₁ ₁
```

```
   364
 ×  87
 29120   364 × 80
  2548   364 × 7
 31668
```

Calculations

Division

Whole numbers
Example: 337 ÷ 8

```
8) 337
  − 80    (8 × 10)
   257
  − 80    (8 × 10)
   177
  − 80    (8 × 10)
    97
  − 80    (8 × 10)
    17
  − 16    (8 ×  2)
     1        42
```

Answer   42 R 1

→

```
8) 337
  − 320    (8 × 40)
    17
  − 16    (8 ×  2)
     1        42
```

Answer   42 R 1

→

```
   42
8) 337
   32
   17
   16
    1
```

→

```
   42  R 1
8) 337
```

Decimals

Example: 78.3 ÷ 9

```
9) 78.3
 − 72.0    (9 × 8)
    6.3
 −  6.3    (9 × 0.7)
    0         8.7)
```

Answer   8.7

Example: 48.6 ÷ 3

```
3) 48.6
 − 30.0    (3 × 10)
   18.6
 − 18.0    (3 ×  6)
    0.6
 −  0.6    (3 × 0.2)
    0        16.2
```

Answer   16.2

Order of operations

Brackets ➡ Division ➡ Multiplication ➡ Addition ➡ Subtraction

Shape and space

2–D shapes

circle   semi-circle   right-angled triangle   equilateral triangle   isosceles triangle   scalene triangle   square   rectangle

rhombus   kite   parallelogram   trapezium   pentagon   hexagon   heptagon   octagon

Shape and space

3–D solids

cube

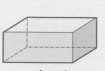
cuboid

cone

cylinder

sphere

hemi-sphere

triangular prism

triangular-based pyramid (tetrahedron)

square-based pyramid

octahedron

dodecahedron

Co-ordinates

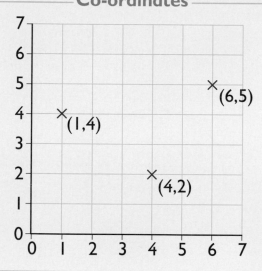

Reflection

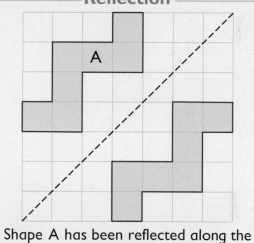

Shape A has been reflected along the diagonal line of symmetry

Rotation

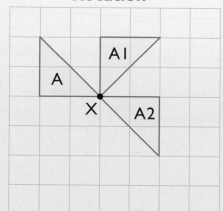

Shape A has been rotated through 90° (Shape A1) and 180° (Shape A2) around Point X

Translation

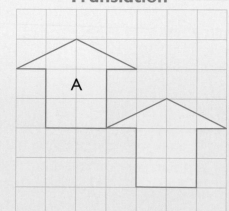

Shape A has been translated 3 squares to the right and 2 squares down.

Shape and space

Angles

Acute angle < 90°
Obtuse angle > 90° and < 180°
Reflex angle > 180° and < 360°
4 right angles (complete turn) = 360°

Right angle = 90°
Straight angle = 180°

Lines

Parallel lines

Perpendicular lines

Measures

Length

1 km	=	1000 m	=	100 000 cm		
0·1 km	=	100 m	=	10 000 cm	=	100 000 mm
0·01 km	=	10 m	=	1000 cm	=	10 000 mm
1 m	=	100 cm	=	1000 mm		
0·1 m	=	10 cm	=	100 mm		
0·01 m	=	1 cm	=	10 mm		
1 cm	=	10 mm	0·1 cm	=	1 mm	

Mass

1 t	=	1000 kg	1 kg	=	1000 g
0.1 kg	=	100 g	0.01 kg	=	10 g

Capacity

1 litre	=	1000 ml	0.1 l	=	100 ml
0.01 l	=	10 ml	1 cl	=	10 ml

Metric units and imperial units

Length	*Mass*	*Capacity*
8 km ≈ 5 miles (1 mile ≈ 1.6 km)	1 kg ≈ 2.2 lb 30 g ≈ 1 oz	1 litre ≈ $1\frac{3}{4}$ pints 4.5 litres ≈ 8 pints (1 gallon)

Time

1 millennium	=	1000 years
1 century	=	100 years
1 decade	=	10 years
1 year	=	12 months
	=	365 days
	=	366 days (leap year)
1 week (wk)	=	7 days
1 day	=	24 hours
1 minute (min)	=	60 seconds

24 hour time

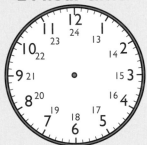

Perimeter and Area

P = perimeter   A = area   l = length   b = breadth

Perimeter of a rectangle:
P = 2l + 2b *or* P = 2 × (l + b)

Perimeter of a square:
P = 4 × l

Area of a rectangle:
A = l × b

Handling data

Planning an investigation

❶ Describe your investigation. ❷ Do you have a prediction? ❸ Describe the data you need to collect.
❹ How will you record and organise the data? ❺ What diagrams will you use to illustrate the data?
❻ What statistics will you calculate? ❼ How will you analyse the data and come to a conclusion?
❽ When you have finished, describe how your investigation could be improved.

Mode	**Range**	**Median**	**Mean**
The value that occurs most often.	Difference between the largest value and the smallest value.	Middle value when all the values have been ordered smallest to largest.	Total of all the values divided by the number of values.